John G. Lenard (Editor)

Modelling Hot Deformation of Steels

An Approach to Understanding and Behaviour

Springer-Verlag Berlin Heidelberg NewYork
London Paris Tokyo

Professor John G. Lenard

Dept. of Mechanical Engineering
University of New Brunswick
Fredericton, N.B., E3B 5A3
Canada

ISBN 3-540-50754-X Springer-Verlag Berlin Heidelberg New York
ISBN 0-387-50754-X Springer-Verlag New York Berlin Heidelberg

Library of Congress Cataloging-in-Publication Data
Modelling hot deformation of steels:
an approach to understanding and behavior / John G. Lenard, editor.
Includes bibliographies and indexes.
ISBN 0-387-50754-X (U.S.)
1. Metals--Hot working--Computer simulation.
I. Lenard, John G.
TS209.5.M63 1989
672.3--dc19 89-4174

Printed in Germany

Offsetprinting: Color-Druck Dorfi GmbH, Berlin
Bookbinding: Lüderitz & Bauer, Berlin

2161/3020 543210 Printed on acid-free paper

PREFACE

Computer Aided Engineering may be defined as an approach to solving technological problems in which most or all of the steps involved are automated through the use of computers, data bases and mathematical models. The success of this approach, considering hot forming, is tied very directly to an understanding of material behaviour when subjected to deformation at high temperatures. There is general agreement among engineers that not enough is known about that topic - and this gave the initial impetus for the project described in the present study. The authors secured a research grant from NATO (Special Research Grant #390/83) with a mandate to study the "State-of-the-Art of Controlled Rolling". What follows is the result of that study.

There are five chapters in this Monograph. The first one, entitled "State-of-the-Art of Controlled Rolling" discusses industrial and laboratory practices and research designed to aid in the development of microalloyed steels of superior quality. Following this is the chapter "Methods of Determining Stress-Strain Curves at Elevated Temperatures". The central concern here is the material's resistance to deformation or in other words, its flow strength, the knowledge of which is absolutely essential for the efficient and economical utilization of the computers controlling the rolling process. "Metallurgical Study of the Hot Upsetting of 1035 Steel" follows. In that the author focusses attention on dynamic recovery and recrystallization during testing for strength. In "Computer-Aided Analysis and Modelling of Plastic Behaviour of Steels at Elevated Temperatures" the representation of constitutive data, using advanced non-linear regression analysis techniques is discussed.

In the last chapter, "Mapping Dynamic Material Behaviour", a new technique is described which allows the engineer to decide on process parameters that ensure efficient operations.

In the Appendix some experimental results concerning the hot strength of two niobium bearing microalloyed steels are presented.

As with any review, this one does not pretend to be complete. Many worthy and significant contributions must have been overlooked and for this we apologize. It is hoped that the present work will further discussions regarding the metal's behaviour during plastic working.

The Authors are grateful to Dr. T. Tanaka, the American Society for Metals and the Institute of Metals for permission to reproduce Figure 4 of "Controlled Rolling of Steel Plate and Strip" (International Metals Reviews, ASM International, Metals Park, OH 44073, USA, 1981, No. 4); to the Metallurgical Society for permission to reproduce Figure 1 of "Ferrite Formation from Thermo-Mechanically Processed Austenite" (R.K. Amin and F.B. Pickering, Thermomechanical Processing of Microalloyed Austenite, edited by A.J. DeArdo, G.A. Ratz and P.J. Wray, 1982, The Metallurgical Society, 420 Commonwealth Drive, Warrendale, Pennsylvania 15088) and to the Institute of Metals for permission to reproduce Figures from "The Torsion Test - Plastic Deformation to High Strains and High Strain Rates" by Pöhlandt and Tekkaya, Material Science Technology, Vol. 1, 1985.

Thanks are also due to Prof. Ashby of Cambridge University for his permission to reproduce the deformation map of copper, first published in a Cambridge University Engineering Department Report; to the MIT Press for use of Fig. 4.1, p. 129 of "Constitutive Equations in Plasticity", edited by A. Argon and to ASM International for permission to reproduce Figures 2 to 8 and 10 to 15 from "On the Response of Nb Bearing HSLA Steels to Single and Multistage Compression" by D'Orazio, Mitchell and Lenard (HSLA Steels-Metallurgy and Applications, Beijing, 1985).

The Authors wish to express their gratitude to NATO for providing financial help during the tenure of this project. The editor would like to express his appreciation to Professors Alexander, Kaftanoglu, Lange and von Turkovich for their splendid cooperation, stimulating discussions and their contributions. Further, the assistance of Muriel Sullivan and Susan Shea during the preparation of this monograph needs to be acknowledged.

J.G. Lenard
Editor
Spring, 1989

TABLE OF CONTENTS

Chapter 3
METALLURGICAL STUDY OF THE HOT UPSETTING OF 1035 STEEL

Chapter 4
COMPUTER-AIDED ANALYSIS AND MODELLING OF PLASTIC BEHAVIOUR OF STEELS AT ELEVATED TEMPERATURES

Chapter 5

MAPPING DYNAMIC MATERIAL BEHAVIOUR

Appendix

Chapter 1

State-of-The-Art of Controlled Rolling

by J.G.Lenard

1. ABSTRACT

A discussion of published research of controlled rolling of steels is presented. Simulation of controlled rolling and industrial practice are reviewed. Some general conclusions are drawn.

2. INTRODUCTION

Conventional hot rolling of plates, strips, bars, rods and shapes has usually been performed at as high temperatures as possible, thereby reducing the load on the work rolls and mill stands as well as contributing to increased productivity. Alloying with various elements, such as carbon, manganese, chromium, nickel etc., was considered when increased strengths were sought. Customer demands for even higher productivity, lower costs and lower weight/strength ratios led to the development of different grades of steels, among them the family of microalloyed steels. As well, new techniques of processing have evolved, most notable being the combination of heat treatment and mechanical processing during rolling. LIU [1] has identified VANDERBECK [2] to have coined the term "controlled rolling" to identify the above process.

MCQUEEN and JONAS [3] define controlled rolling as a process during which the effects of hot working and of the rate of cooling on the characteristics of recrystallization are utilized, with the objective of achieving a fine, uniform grain size. As described, metallurgical and mechanical properties of the resulting product depend almost exclusively on the manner of plastic working, temperature and rate of deformation, or in other words, on the thermal-mechanical treatment.

Research and development work concerning controlled rolling has been proceeding along several distinct but interconnected avenues. These include detailed studies of the metallurgy of hot working and mechanical testing to establish flow curves using various methods such as tension, torsion and compression. Laboratory simulation of the processes during the roughing and/or finishing stages of strip rolling by multistage testing and multipass rolling on laboratory or full scale mills have also been reported.

Two excellent reviews of the controlled rolling process have been published recently. One is due to ROBERTS [4] while and the other is by TANAKA [5]. As well, the discussion of dynamic microstructural changes during hot deformation by TANAKA et al. [6] needs to be noted.

In what follows these reviews are presented briefly along with their conclusions. Then, some recent work is discussed, focussing on aspects of physical metallurgy of hot rolling, laboratory simulation of controlled rolling, experimental controlled rolling and industrial practice. As is the case with most reviews, the present one will not pretend to be complete.

3. THE REVIEW OF TANAKA [5]

Following a brief description of the history of research concerned with controlled rolling, TANAKA defines the major purpose of the process as ... "to refine the structure of the steel and thereby to enhance both strength and toughness". He discusses the three stages of controlled rolling, proposed in [6], [7] and [8], as

i) deformation in the γ recrystallization region;
ii) deformation in the non-recrystallization region;
iii) deformation in the $\gamma-\alpha$ two-phase region.

As shown in Figure1(reproduced from Reference[5]),coarse austenite, designated "a",is refined by repeated recrystallization and deformation.In Stage 2 deformation bands are formed in the elongated,unrecrystallized austenite and ferrite nucleates on the deformation bands and on the austenite grain boundaries.The deformation in the $\gamma-\alpha$ two phase region continues in Stage 3 producing a substructure. During cooling unrecrystallized austenite transforms to equiaxed α grains and the ferrite changes into subgrains.

The difference between controlled and conventional rolling is then observed from the above discussion and Figure 1. In conventional rolling ferrite grains nucleate on the γ grain boundaries only while in controlled rolling nucleation occurs at grain interiors as well.

In order for dynamic recrystallization to begin a critical strain level must be experienced by the rolled steel. Due to the retarding effect of the microalloyed

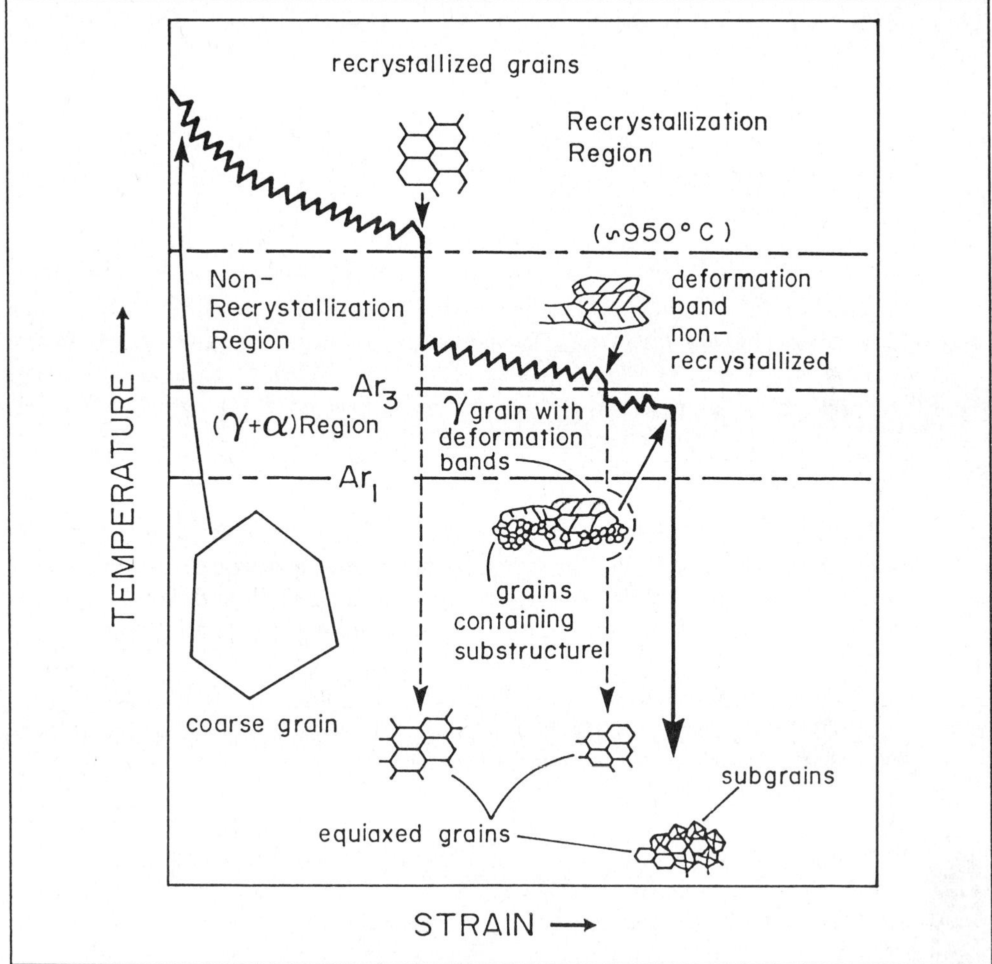

Figure 1: Schematic illustration of three stages of controlled rolling process and change in microstructure with deformation in each stage;after [5].

elements this critical strain is fairly large and hence it is almost impossible to hot roll HSLA steel in the dynamically recrystallized, steady state regime. Therefore, grain refinement must proceed by static recrystallization which does not occur under all conditions.

The grain size may be decreased further by deformation in the nonrecrystallized region. This refinement reaches a limiting value at 60-70% reduction.

In addition to the presentation of the metallurgical phenomena accompanying controlled rolling, TANAKA also discusses the properties of controlled rolled steels, including anisotropy and embrittlement. As well, he presents empirical equations giving the strength and toughness of the steels.

When writing about the effect of alloying elements, he considers Nb, Ti and V in some detail. Si, Mn, Cr, Cu, Ni and Mo are also mentioned briefly. Niobium contributes to strength by grain refinement and precipitation; vanadium increases both strength and the transition temperature through precipitation hardening. The effect of titanium in strengthening is less than that of Nb and of V. As well, Ti decreases the toughness of the steel and of the heat affected zone; TANAKA hence concludes that its use is limited.

In industrial practice, the important factors to be taken into account are listed by TANAKA as the

i) lowering of the slab reheating temperature
ii) grain refinement via recrystallization
iii) selection of pass schedules.

4. THE REVIEW OF ROBERTS [4]

In his recent review, ROBERTS focusses on advances in alloy design and processing technology of microalloyed steels. He lists the significant events in the progress in steel technology as

i) new steel making procedures providing for low levels of impurities;
ii) continuous casting technology;
iii) novel alloying and processing technology;
iv) improvements in weldability, formability and resistance to stress corrosion of HSLA steels;
v) accelerated or controlled cooling.

In discussing alloy design he examines the role of titanium in some detail.As he writes, Ti additions are often made to control grain sizes.

The advantages of microalloying with titanium include its ability to control austenite grain growth and exceptional heat-affected-zone grain coarsening resistance. To achieve maximum degree of precipitation strengthening careful control of the coiling temperature is essential.

ROBERTS then discusses the advantages and limitations of boron heat treatable steels. The advantages are:

- low as-rolled hardness
- good cost effectiveness
- possibility of direct hardening without cracking.

The disadvantages include limited hardenability, higher hardening temperature and poorer temper resistance. ROBERTS' conclusions concerning microalloy design are worth repeating here. The major advances include:

i) development of HSLA steels containing small amounts of Ti whose strength is increased by TiC precipitation;
ii) development of ultra low carbon bainitic steels and heat treatable boron alloy grades with consistent hardenability;
iii) development of V-microalloyed steels with high N content;
iv) development of medium carbon QT steels and ferrite/pearlite steels with V and/or Nb.

The author's conclusions regarding the processing technology of microalloyed steels list the major, significant innovations as the

i) introduction of accelerated cooling, interpass or interstand cooling and
ii) recrystallization controlled rolling.

5. PHYSICAL METALLURGY OF HOT ROLLING

The principles of the physical metallurgy of hot working have been described succinctly by SELLARS [9] in his keynote address to the Sheffield, 1979 conference. As he wrote, work hardening during deformation is balanced by the dynamic softening processes of recovery and recrystallization. The microscopic changes result in an increase of dislocation density until a critical strain level is reached and the stored energy is sufficiently high to cause the beginning of dynamic recrystallization.

Further straining causes repeated dynamic recrystallization as the new grains are themselves work hardened.

These changes leave the metal in an unstable state and provide the driving force for static recovery and static recrystallization, which may be followed by grain growth.

SELLARS also writes that it is unlikely that the critical strains will be passed in plate rolling but that they will be exceeded during strip rolling.

As he states, answers to two questions are required in order to apply the above principles in commercial processes. These are

a) How long does recrystallization take after deformation?
b) What grain size is produced by the above events?

In a more recent publication SELLARS [10] has summarized the options available for thermomechanical processing, which depend on the solubility of carbonitrides. As he states, during controlled rolling of niobium steels precipitation begins during roughing. During finishing strain induced precipitation retards recrystallization and can also lead to significant precipitation strengthening. When titanium steels are considered fine recrystallized austenite grains can be obtained without low finish rolling temperatures. Vanadium does not reprecipitate until transformation and hence it has a minor influence on rolling conditions.

As JONAS [11] writes, full recrystallization is to occur in the roughing mill but it is to be avoided during controlled rolling in the finishing stands. In recrystallization controlled rolling the strain free grains are to form in between passes from one mill stand to another.

6. SIMULATION OF CONTROLLED ROLLING

As MCQUEEN [12] in his review of simulation of multistage hot-forming writes, the sequential processes of forming and rolling have many variables that require control. These include the temperature, rate of temperature change, strain and strain rate in each of the cycles. As well, these variables must also be monitored and/or controlled during the interpass periods.

Tension, compression and torsion tests are used for the simulation of the finish rolling process. Tension testing, often done on a Gleeble-testing machine [13] - [15], is suitable if strains are not to exceed 15-20%. Compression may be used to produce logarithmic strains of 1.5-2, see Rf. [16] to [20], while torsion may be carried to a total strain of 20, [21]-[24]. In the opinion of the present writer, compression testing is probably best suited for simulation of rolling which is an essentially compressive process [25].

During testing the process parameters must be carefully controlled. On camplastometers [16] the cams need to be carefully designed while on microprocessor-controlled servo-hydraulic testing systems [19]-[20] the parameters may be entered interactively, on the computer's keyboard. It is the computer controlled servohydraulic tester that affords the most ease of handling and accuracy.

WEISS et al. [25] have investigated high temperature recrystallization through the use of the hot compression test. Several V-N and V-Nb-N steels were used. The compression tests were carried out after solution treatment at 1200°C for 30 minutes, followed by air cooling to test temperatures. After testing the samples were quenched in ice water.

The authors determined the true strains required to reach peak stress levels, beyond which, softening due to dynamic recrystallization is indicated by a fall in the flow curve.

Experiments at higher constant strain rates - at 2 s^{-1} and 13 s^{-1} - indicated that both Nb and V cause an increase in time parameters, defined as

$$t_p = \varepsilon_p / \dot{\varepsilon}$$

In general, the delay of recrystallization was only weakly dependent on temperature and strain rate. The writers agree with the conclusion of AMIN and PICKERING [26] that this delay is caused by solute drag effects. At low strain rates, however, retardation of dynamic recrystallization became strongly temperature dependent.

KNUDSEN et al. [22] have used interrupted torsion testing to simulate the rolling schedules of a reversing hot mill with hot coilers, used by SIDBEC-DOSCO (Canada). Niobium bearing steel, containing 0.05% Nb was used and up to 15 interruptions with various interpass times (10 to 150 seconds) were employed. Strain rates of up to 10 s^{-1} were reached. The authors confirmed that niobium impedes the progress of recrystallization and reduces the rate of static restoration. As well, they found that at low strains, low strain rates and delay times not exceeding 40 seconds static recovery is the significant softening mechanism, while for higher rates, higher strains or larger delay times static recrystallization took place. Metadynamic recrystallization was found to occur for strains reaching the steady state regime.

EVERETT et al. [24] have investigated the high temperature strengths of some 22 steels, containing various amounts of niobium and/or vanadium and/or silicon. Using a hot torsion tester, they attempted to simulate the conditions existing during hot strip rolling. The first four stands of a finishing train were considered; processing temperatures of 875-1050°C were used and a constant strain rate of $7s^{-1}$ was employed.

The authors concluded that the retardation of softening is highest with steels containing Nb and somewhat less with vanadium or solid solution hardened steels.

The resistance to deformation of eight low carbon steels containing various amounts of niobium (0-0.1%) was measured in interrupted torsion testing by MIGAUD [21]. Using a temperature range of 900°C to 1150°C, he developed empirical relations for the flow strength in terms of the strain, rate of strain and the torque. He concludes that the rate of strain - at least in the range of his experiments - is not a significant parameter as far as the simulation experiments are concerned.

The practice of using multistage, uniaxial testing to simulate sequential hot forming processes has led to a better understanding of those processes. Microstructures, resulting from simulation or from actual rolling tests, have been shown to be similar, providing that parameters - strain, strain rate and temperature - were identical [19]. Difficulties, however, exist and in the opinion of the present writer, listing and discussing them may be of some benefit.

One of the most important parameters in hot testing is the temperature. It is well known that at 1000°C a 5% variation of temperature may well lead to a 25-40% change of strength. Still, review of the technical literature reveals that not enough attention has been paid to the control and/or measurement of temperature. In some cases thermocouples, in others, optical pyrometers are used to monitor the test temperature. In References [23] and [27] the thermocouples made direct contact with the samples. In Reference [16] corrections for adiabatic heating were included. In many other instances the test temperature is simply stated with no explanation regarding the method of measurement or any corrections. When these uncertainties are combined with the accuracy of a chromel-alumel type thermocouple (usually 3/4% full scale) or with the lack of precise knowledge of the dependence of emissivity on temperature, doubts arise as to the validity of published strength data.

Of equal importance for microalloyed steels is the pre-test thermal treatment the samples are subjected to. As noticed in References [16], [23], [27] and [28] all dealing with Nb bearing HSLA steels, solution temperatures vary from 1100°C to 1300°C; times vary from 1 to 60 minutes. Probably as a result of both of the above discussed matters, for 0.038% Nb bearing steel TIITTO et al. [29] and MAKI et al. [30] present data that differ by approximately 100%.

Other parameters, such as the dynamics of the testing machine and effectiveness of lubrication also influence the results.

It appears that standardization of the hot compression test, including procedures, measurement techniques and the manner of reporting the results should be considered by the researchers.

7. CONTROLLED ROLLING - LABORATORY MILLS

AMIN and PICKERING [31] have used controlled rolling to examine its effect on the recrystallization of austenite. Some 18 steels, containing Nb, N-V and C,N and Nb were used. After solution treating all steels at 1300°C, hot rolling was carried out according to the schedules shown in Figure 2.

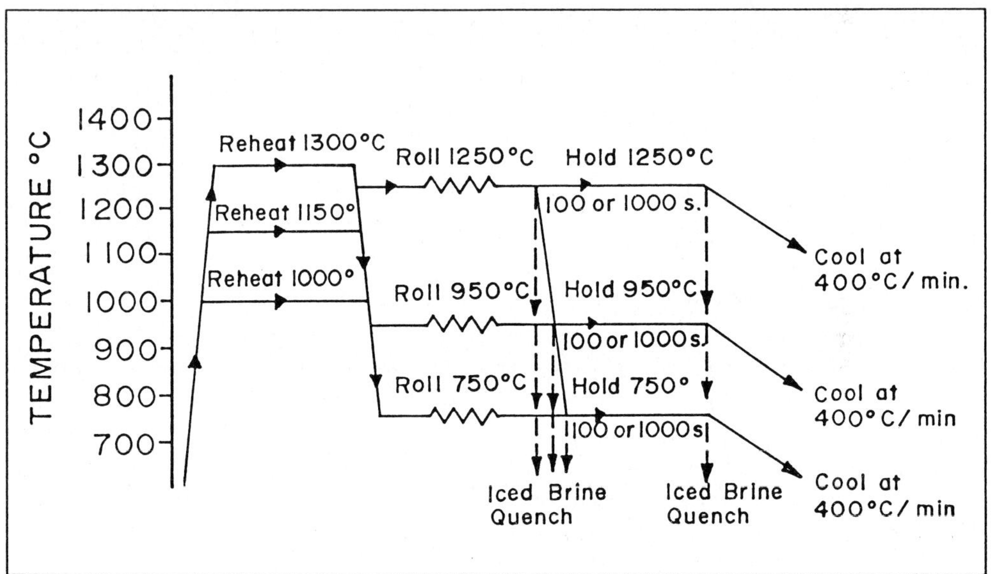

Figure 2: Thermal-mechanical treatment of reference [31].

They found that increasing Nb content refined the austenite grain size. A limiting size was reached however, below which even the highest Nb content (1.03%) did not produce further refinement. Further, they show that increasing the reheating temperature will increase the Nb in solution which in turn leads to retardation of recrystallization. The rolling temperature for complete recrystallization increased with increasing Nb content, with decreasing rolling reduction and with increasing reheating temperature prior to processing. As well, in examining the effects of vanadium on recrystallization, the authors write that dissolved V is not very effective in retarding recrystallization but that precipitation of VN at 950°C and below had an inhibiting effect. Nb was shown to have a higher grain coarsening effect than V. Retardation of recrystallization was caused by a solute drag effect at high temperatures while at lower temperatures strain induced precipitation inhibited nucleation and growth of the recrystallizing grains.

SANTELLA and DEARDO [32] performed hot rolling experiments at mean strain rates of 3.8 s^{-1} in order to investigate the possibility of alternate methods of producing refined grains. They found that the recrystallized grain size was a function only of the rolling temperature and that Nb(CN) particles enhanced the nucleation of recrystallization accompanying rolling.

DIMICCO and DAVENPORT [33] used multipass commercial rolling schedules to examine their effects on the structural changes of austenite. They used bainitic steels and, for comparison, C-Mn steels as well. They concluded that increasing total reduction, increasing reduction/pass and decreasing rolling temperature contributed to the refinement of austenite grain sizes. The most important factor controlling grain refinement was found to be the total reduction.

They agreed with [28] in concluding that Nb additions up to 0.1% have a strong grain refining effect. The bainitic steels had a larger grain size than steels carrying Nb only. Further they confirmed the existence of strain induced abnormal grain growths and recommended that reductions/pass be above 10-14% to avoid it.

ORRLING, SJOSTROM and HEEDMAN [34] used a pilot plant mill to study the technique of controlled rolling and direct quenching of plates. Seven different steels, containing vanadium and/or niobium were rolled. Their objective of reducing process time was based on the observation of a relatively long holding period and more passes in the finishing stages required by HSLA steel. They concluded that several ways are available to decrease process time in one stand reversible plate mills. One may use reduced roughing temperatures; another is to roll many plates at the same time; another still is the use of water cooling during either the roughing stage or during the holding period. The authors state that forced cooling during the roughing stage can reduce the process time by 65-75% but results in a somewhat deteriorated toughness.

WARREN, UBHI and JACK [35] carried out controlled rolling experiments to investigate whether replacement of niobium and molybdenum by vanadium would result in significant changes in the mechanical and metallurgical properties of the material. After an extensive study of microstructures the authors concluded that the structures of the two steels are similar. Further, they stated that the structures consist of ferrite, formed during and after final rolling as well as 4 to 11% hard, carbon rich constitutent. As well, vanadium will precipitate as VC, will retard recrystallization and will form a fine grained structure; small amount of molybdenum or accelerated cooling may prevent interphase reactions. Finally, they concluded that a decrease in finish rolling temperature leads to higher strength.

REUBEN and BAKER [36] studied the possibility of developing high strength and good toughness by adding 4% Mn to Nb bearing low carbon steels. They used controlled rolling to produce fine microstructures.

Alloys were subjected to two different solution treatments. In one, the samples were held at 1300°C for one hour and then rolled; in the other heat treatment at 1150°C-1200°C was employed. In the rolling schedule five passes were used, the treatment ending with finish rolling temperatures varying from 700-950°C.

The authors concluded that the addition of niobium has a grain refining effect and leads to improved toughness and strength for finish rolling temperatures above 800°C. Increasing the carbon content also led to an increase in strength. As well, lower solution treatment temperatures had no observable effect on the final properties.

Hot rolling experiments are more difficult to perform than the simulation tests. One of the difficulties is again the measurement and control of the temperature; but now chilling at the roll-strip interface also needs to be considered and this, of course, further complicated procedures. As well, most laboratory mills are of the single stand construction which makes multistage rolling studies extremely cumbersome.

Use of results, obtained on small-scale mills, for prediction of events on industrial scale stands, has been questioned by engineers in the steel industry. A statistical study, comparing the standard deviation of differences between predictions and measurements, obtained in productions runs and in the laboratory, showed some promise [37].

8. CONTROLLED ROLLING - INDUSTRIAL PRACTICE

It is practically impossible to obtain detailed and precise information regarding controlled rolling and recrystallization controlled rolling followed by the steel industry. Understandably chemical compositions and corresponding specific draft schedules are considered proprietary information by the companies. The review writer must then be content with discussion of the articles published in the technical literature.

There appears to be general agreement among steel mill engineers regarding the objectives to be achieved by controlled rolling - that of achieving consistently improved properties, lower costs and higher productivity. The methods to fulfill these objectives at several steel mills will be discussed below.

ACKERT and BOWIE [38] have reported on their experiences with controlled rolling on Algoma Steel Corporation, Ltd.'s automated 4220 mm wide plate mill. The stated objectives were to increase productivity, to improve repeatability of impact properties, to broaden the product range and to produce a wide range of product. They describe commissioning the mill, the original rolling scheme and the modifications performed as a result of rolling Nb bearing HSLA plates. In reading their conclusions it is apparent that their objectives have been met satisfactorily. As the last comment they write: "Rejects due to the system failing to achieve the proper reduction sequence have been non-existent."

In discussing the development of API X80 pipeline steel PONTREMOLI et al. [39] define the major objective: to design the chemical composition such that the material will exhibit a continuous stress-strain curve as well as yield adequate weldability. Controlled rolling schedules, as given below, were used

Prior heating:	1150°C for 2 hours
Finish rolling temperature:	730-750°C
Initial thickness:	110 mm
Roughing reduction:	51%
Finishing reduction:	64%
Final thickness:	18 mm

for three groups of steels. The first group's composition was C-Mn-Nb-Cr; the second group also contained Si and Ni while the third group had much lower carbon content in addition to 0.001 - 0.0025% boron.

The authors concluded that the martensite-austenite constituent was much more finely dispersed in the boron steels. Both boron free and boron steels yielded the required strength but those containing boron showed superior toughness levels. Further, the use of titanium was found to be advantageous in its ability to increase strengthening by precipitation hardening. Mention is also made of the possibility of using accelerated cooling to achieve the same objectives but with less alloying.

In a more recent paper BUFALINI et al. [40] have studied the effect of "start cooling temperature", cooling rate and finishing cooling temperature on the structure and properties of HSLA steels. They found that with their accelerated cooling process, best results were obtained when cooling began above the Ar_3 temperature at a rate of 15°C/ second. The most significant parameter, controlling the final structure was the temperature at which cooling was stopped.

BOER et al. [41] state the requirements for thermomechanical treatment as a powerful rolling mill, comprehensive control system and exact temperature measurements. They also list the essential parameters as the reheating temperature, rolling temperature, reduction per pass and cooling rate after finish rolling.

WILLIAMS et al. [42] describe the steps taken to develop a cost effective process to improve weldability of HSLA plates. Experiments were conducted in the laboratory and in production runs and the objectives were achieved.

9. CONCLUSIONS

A brief review of recent research on controlled rolling has been presented. Simulation of the process by interrupted torsion testing, laboratory and industrial practices has been mentioned.

10. REFERENCES

1. Liu, L., "Simulation of Controlled Rolling in Two Ti HSLA Steels", M.Eng. Thesis, McGill University, Montreal, Canada, 1983

2. Vanderbeck, R.W., Welding Journal, 37, 1958, pp. 1145-1165.

3. McQueen, H.J. and Jonas, J.J., "Hot Workability Testing Techniques", in Metal forming: Interrelation Between Theory and Practice, Editor: A.L. Hoffmanner, Plenum Press, New York, 1971, pp. 393-478.

4. Roberts, W., "Recent Innovations in Alloy Design and Processing of Microalloyed Steels", Proc. HSLA 83, 1983, Philadelphia, Pa., pp. 33-66.

5. Tanaka, T., "Controlled Rolling of Steel Plate and Strip", International Metals Reviews, No. 4, 1981, pp. 185-212.

6. Tanaka, T., Tabata, N., Hatomura, T. & Shiga, C., "Three Stages of the Controlled-Rolling Process", Micro alloying 75, Proc. Int. Symp. on HSLA Steels, Washington, 1975, 107-119.

7. Kozasu, I., Shimizu, T. and Kubota, H., Trans. Iron & Steel Inst. Japan, 1971, 11, pp. 367-375.

8. Baird, J.D. and Preston, R.R., "Processing and properties of low carbon steel", 1-40, 1973, Metallurgical Society of AIME, New York.

9. Sellars, C.M., "The physical metallurgy of hot working", Proc. Int. Conf. on Hot Working Forming Processes, Sheffield, England, 1979, pp. 3-15.

10. Sellars, C.M., "Options and Constraints for Thermomechanical Processing of Microalloyed Steel", Proc. HSLA Steels 85, 1985, Beijing, China, pp. 73-82.

11. Jonas, J.J., "Mechanical Testing for the Study of Austenite Recrystallization and Carbonitride Precipitation", Proc. Int. Conf. on HSLA Steels, 1984, Wollongong, Australia.

12. McQueen H.J., "Review of Simulations of Multistage Hot Forming of Steels", Can. Met. Quarterly, 1982, 21, pp.445-460.

13. Wilber, G.A., Bell, J.R., Bucher, J.H. and Childs, W.J., Trans. Metall. Soc. AIME, 1968, 242, pp. 2305-2308.

14. Capeleth, T.L., Jackman, L.A. and Childs, W.J., Metall. Trans., 1972, 3, pp. 789-796.

15. Schmidtmann, E. and Otto, A., "Simulation der Warm um formung mikro-legierter feinkornbaustahle im Mehrstufer-Zugversuch", Archiv für das Eisenhüttenwesen, 1982, 53, pp. 275-285.

16. Steward, M.J., "Constant True Strain Rate Compression: The Camplastometer", Can. Met. Quart., 1974, 13, p. 503.

17. Luton, M.J., Immarigeon, J.P. and Jonas, J.J., "Constant True Strain Rate Apparatus for use with Instron Testing Machines", J. Physics, 1974, 7, p. 862.

18. Suzuki, H., Hashizume, S., Yabuki, Y. and Ichihara, Y., "Studies on the Flow Stress of Metals and Alloys", University Tokyo Report, 18, 1968, p. 139.

19. Lenard, J.G., "Development of an Experimental Facility for Single and Multistage, Constant Strain Rate Compression", ASME, J. Engng. Mat. & Techn., 107, 1985, p. 126.

20. D'Orazio, L.R., Mitchell, A.B. and Lenard, J.G., "On the Response of Nb Bearing HSLA Steels to Single and Multistage Compression", Trans. HSLA 85, 1985, Beijing, PRC, p. 189.

21. Migaud, B., "Simulation by Hot Torsion Testing of Hot Strip Mill Rolling for Low-Carbon Nb-Microalloyed Steels", Proc. Int. Conf. on Hot Working and Forming Processes, Sheffield, 1979, p. 67.

22. Knudsen, W., Sankar, J., McQueen, H.J., Jonas, J.J. and Hawkins, D.N., "Simulation of Rolling Schedules for HSLA Steels", Proc. Int. Conf. on Hot Working and Forming Processes , Sheffield, 1979, p. 51.

23. Sankar, J., Hawkins, D. and McQueen, H.J., "Behavior of Low-Carbon and HSLA Steels During Torsion - Simulated Continuous and Interrupted Hot-Rolling Practice", Metals Technology, 1979, p. 375.

24. Everett, J.R., Gittins, A., Glover, G. and Toyama, M., "Effects of Microalloys in Hot Strip Mill Rolling", Proc. Int. Conf. on Hot Working and Forming Processes, Sheffield, 1979, p. 16.

25. Weiss, I., Fitzsimons, G.L., Mielityinen-Tiitto, K. and DeArdo, A.J., "The Influence of Niobium, Vanadium and Nitrogen on the Response of Austenite to Reheating and Hot Deformation Microalloyed Steels", Proc. Int. Conf. on Thermomechanical Processing of Microalloyed Austenite, Pittsburgh, 1981, p. 33.

26. Amin, R.K. and Pickering, F.B., "Austenite Grain Coarsening and the Effect of Thermomechanical Processing on Austenite Recrystallization", Proc. Int. Conf. on Thermomechanical Processing of Microalloyed Austenite, Pittsburgh, 1981, p. 1.

27. Wilcox, J.R. and Honeycombe, R.W.K., "Hot Ductility of Nb and Al Microalloyed Steels Following High-Temperature Solution Treatment", Met. Technology, 11, 1984, p. 217.

28. Gittins, A., Everett, J. R. and Tegart, W.J. McG., "Strength of Steels in Hot Strip Mill Rolling", Met. Technology, 1977, p. 377.

29. Tiitto, K., Fitzsimons, G.L. and DeArdo, A.J., "The Effect of Dynamic Precipitation and Recrystallization on the Hot Flow Behavior of a Nb-V Microalloyed Steel", Acta Metall., 31, 1983, p. 1159.

30. Maki, T., Akasaka, K. and Tamura, I., "Dynamic Recrystallization Behavior of Austenite in Several Low and High Alloy Steels", Proc. Int. Conf. on Thermomechanical Processing of Microalloyed Austenite, Pittsburgh, 1981, p. 217.

31. Amin, R.K. and Pickering, F.B., "Ferrite Formation from Thermomechanically Processed Austenite", Proc. Int. Conf. on Thermomechanical Processing of Microalloyed Austenite, Pittsburgh, 1981, p. 377.

32. Santella, M.L. and DeArdo, A.J., "The Hot-Rolling Behavior of Austenite in Nb-Bearing Steels: The Influence pf Reheated Microstructure", Proc. Int. Conf.on Thermomechanical Processing of Microalloyed Austenite, Pittsburgh, 1981, p. 83.

33. DiMicco, D.R. and Davenport, A.T., "Austenite Recrystallization and Grain Growth During the Hot Rolling of Microalloyed Steels", Proc. Int. Conf. on Thermomechanical Processing of Microalloyed Austenite, Pittsburgh, 1981, p. 59.

34. Orrling, B., Sjostrom, A. and Heedman, P., "Controlled rolling and direct quenching of plate in a pilot plant mill", Proc. Int. Conf. on Steel Rolling, ISIJ, Tokyo, Vol. 2, 1980, p. 909.

35. Warren, A., Ubhi, H.S. and Jack, D.H., "Relationship between processing, microstructure and properties in controlled rolled 0.45% V steel", Metal Science, Vol. 17, 1983, p. 19.

36. Reuben, R.L. and Baker, T.N., "Mechanical properties and microstructure of low carbon, 4% manganese steels", Metals Technology, 11, 1984, p.6.

37. Murthy, A. and Lenard, J.G., "Statistical Evaluation of Some Hot Rolling Theories", ASME, J. Engng. Mat. & Techn., 104, 1982, p. 47.

38. Ackert, R.J. and Bowie, M., "Temperature Controlled Rolling of Plate as Developed and Practiced at the Algoma Steel 4220 mm (166 inch) Wide Plate Mill", Proc. Int. Conf. on Steel Rolling, Vol. II, ISIJ, Tokyo, 1980, p. 933.

39. Pontremoli, M., Bufalini, P., Aprile, A. and Jannone, C., "Development of Grade API X80 Pipeline Steel Plates Produced by Controlled Rolling", Met. Technology, 11, 1984, p. 506.

40. Bufalini, P., Buzzichelli, G., Pontremoli, M., Aprile, A., Jannone, C. & Pozzi, A., "Development of High Strength Steels for Structural and Line Pipe Applications Through Highly Controlled Process", Proc. HSLA 85, Beijing, PRC, 1985, p. 457.

41. deBoer, H., Beenken, P., Branscheid, F., Flosdorf, W.,Haumann, W.,

Hof,W. and Litzke, H., "Thermo-mechanical Treatment of HSLA Steels; Fundamentals and Realization", Proc. HSLA 85, Beijing, PRC, 1985, p. 539.

42. Williams, J.G., Killmore, C.R., Williams, I.D. & Wood, J.A., "Production of Tough HSLA Heavy Plates with a Moderate Controlled Rolling Practice", Proc. HSLA 85, Beijing, PRC, 1985, p. 567.

Chapter 2

Methods of Determining Stress-Strain Curves at Elevated Temperatures

by K.Pöhlandt, submitted by K.Lange

1. ABSTRACT

Determination of flow curves by tension, torsion and compression testing is discussed. The difficulties, advantages and disadvantages associated with each method are described.

LIST OF SYMBOLS

a	radius
a;b	dimensions of plane strain upsetting test-piece
A	cross-section of specimen
A_{min}	minimum cross section
C	constant in Equation 12
d	average grain diameter
ε	equivalent strain
F	force
ϕ	natural strain (true strain)
$\dot{\phi}$	strain rate (time derivative of ϕ)
ϕ_u	strain at uniform elongation in tensile test
γ_r	shear strain in torsion test-piece

γ_a	shear strain at the surface of torsion test-piece
γ_p	shear strain at the "critical radius" r_p in torsion specimen
h	height of upsetting specimen
h_0	initial height of specimen
K	constant in Equation 7
k_R	resistance to deformation
r_0	initial length of tensile test specimen
l	length of torsion test-piece
M	torque
m	strain-rate sensitivity index
n	strain hardening coefficient
μ	coefficient of friction
ψ	twisting angle in torsion test
r	instantaneous radius of upsetting test-piece
r_0	initial radius of upsetting test-piece
r	distance from the axis of a torsion test-piece
r_p	"critical radius" in torsion test-piece
ρ	radius of necking in tensile test
s	reduction of height of upsetting test-piece
σ_f	flow stress
σ_{f1}	constant in Equation 8
σ_k	grain size-independent part of yield stress according to Equation 7
τ_p	shear stress at the "critical radius" r_p in torsion specimen

2. INTRODUCTION AND BASIC CONCEPTS

For calculating the forces required for metal forming processes it is necessary to know the stress-strain curves (flow curves) of the metals to be formed. The following text which has been widely adopted from Ref. [1] gives a brief outline of the basic concepts and a description of the most important methods of determining flow curves with special emphasis on elevated temperatures.The flow curve is given by the flow stress as a function of strain where stress is defined by

$$\sigma_f = \frac{F}{A} \tag{1}$$

F being the actual force, and A the actual cross-section of an uniaxially deformed tensile test specimen. The natural strain (true strain) is given by

$$\Phi = \ln\left(1 + \frac{\Delta l}{l_0}\right) \tag{2}$$

where l_0 is the initial length of the test-piece, and Δl is the elongation. Assuming constant volume, the actual cross-section A in Equation 1 is given by

$$A = \frac{\Pi r_0^2 l_0}{l_0 + \Delta l} \tag{3}$$

The function

$$\sigma_f = \sigma_f(\Phi) \tag{4}$$

is called the flow curve. It gives the stress required for plastic deformation under a uniaxial state of stress. The relation between this state and a multiaxial state of stress is described by the concepts of equivalent stress σ and equivalent strain ε [1] . For the uniaxial tensile test, however, strain and equivalent strain are equal, i.e.

$$\varepsilon = \Phi \tag{5}$$

In the following text, for simplicity, the symbol ϕ shall be used if this does not cause misunderstanding. Stress does not only depend on strain but also on strain rate, temperature, and, in general, on directional effects caused by anisotropy [23]. Furthermore, there is a weak dependence on hydrostatic stress σ_m [2]. If the flow curve is known, the ideal work required for metal forming per unit volume can be calculated, i.e.

$$\frac{W_{id}}{V} = \int_0^{\Phi} \sigma_f(\Phi')d\Phi' \tag{6}$$

where the variable of integration has been marked by "'".The shapes of flow curves which have been obtained from experiments can be explained qualitatively by the theory of dislocations [1].

The flow curves of polycrystals deviate from those of monocrystals because the individual grains cannot be deformed independently of the grains in their vicinity. Therefore several "glide systems" [1] have to be actuated in each grain. In many cases the yield point of a polycrystal as a function of grain size can be described by the "HALL-PETCH-equation" [1]

$$\sigma = \sigma_K + \frac{K}{\sqrt{d}} \tag{7}$$

This equation is well fulfilled for many metals, see Figure 1. In the range of high strains, however, the effect of grain size on flow stress is more complex.

From the viewpoint of metal physics, room temperature is not a distinguished value with respect to the elementary processes of plastic deformation. However, for practical reasons, cold and hot forming are defined as deformation at room temperature and deformation at elevated temperatures, respectively. Figure 2 shows the flow curves of some metals which at room temperature are not affected by thermally activated processes.

In such cases the flow stress increases monotonically with growing strain, the slope $d\sigma_f/d\phi$ of the curve being highest at low strains.At elevated temperatures when

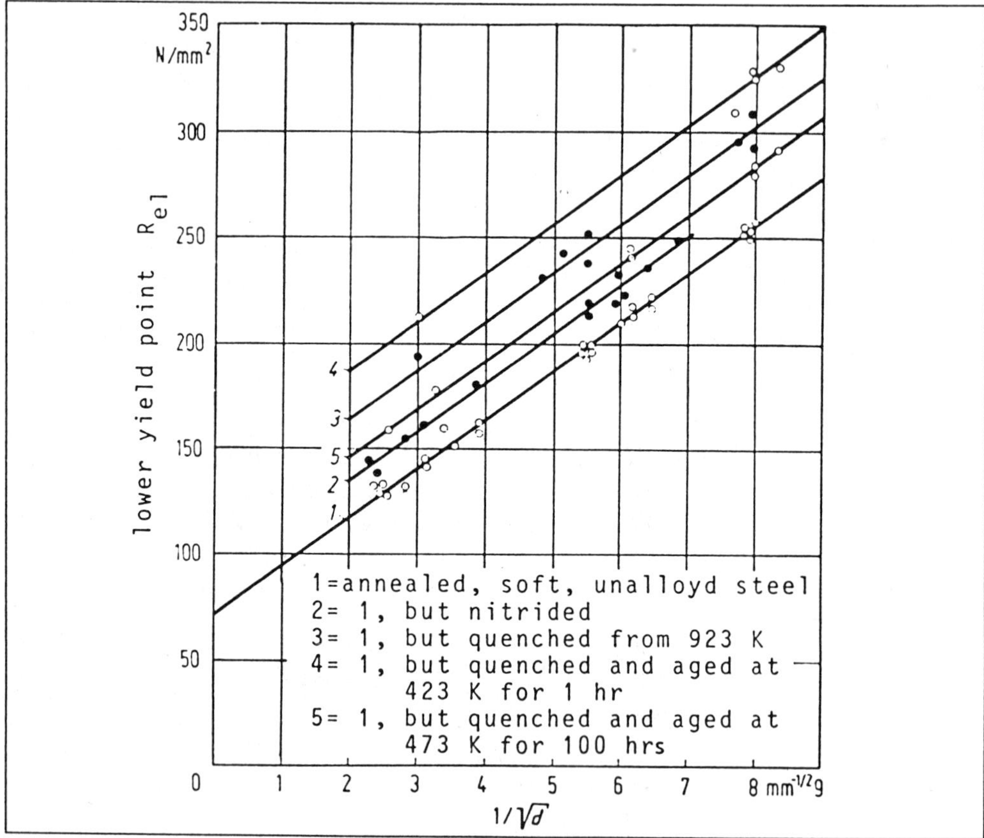

Figure 1: Lower yield point vs. average grain size [4]

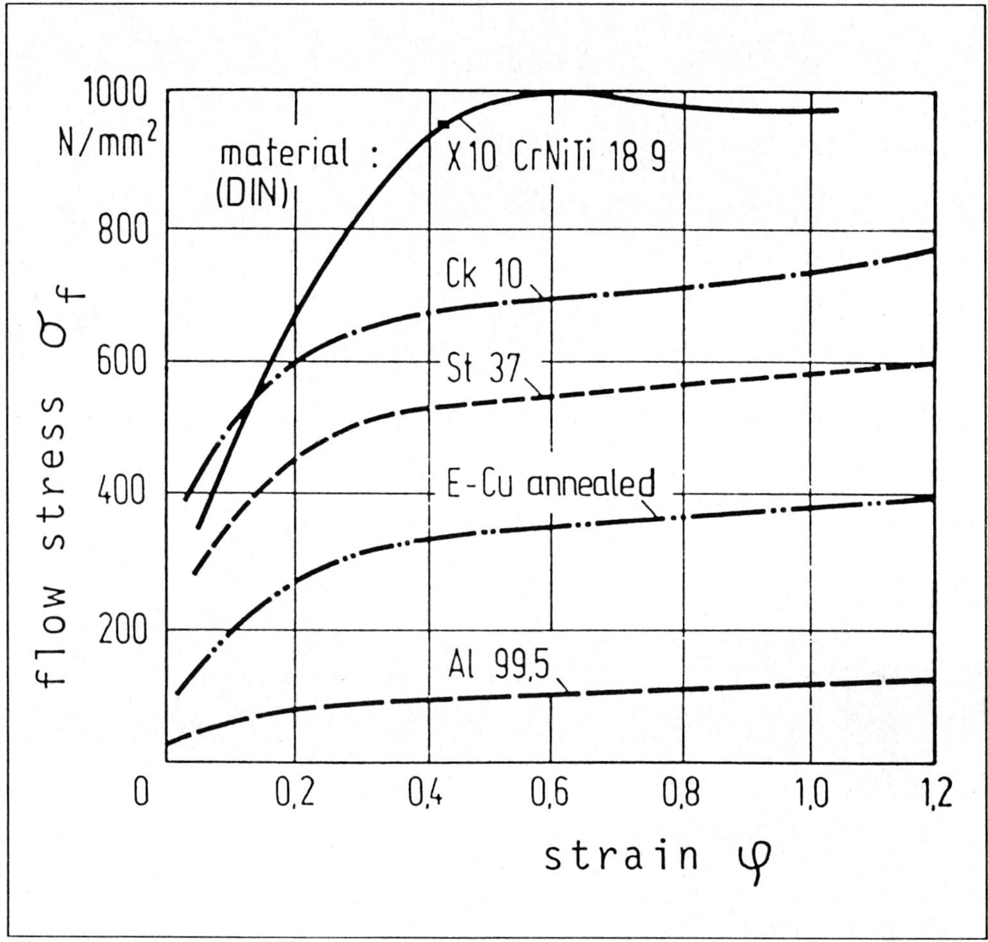

Figure 2: Flow curves of some metals at room temperatures [5] .

recovery and recrystallization [1] take place, the flow curves depend strongly on temperature and strain rate.

Since both recovery and recrystallization take place at a finite, temperature-dependent rate, the flow stress depends strongly on strain rate. At a given temperature, the effect of strain rate on the flow stress can be approximated by the relation

$$\sigma_f \cong \sigma_{f1}\left(\frac{\dot{\Phi}}{\dot{\Phi}_1}\right)^m \tag{8}$$

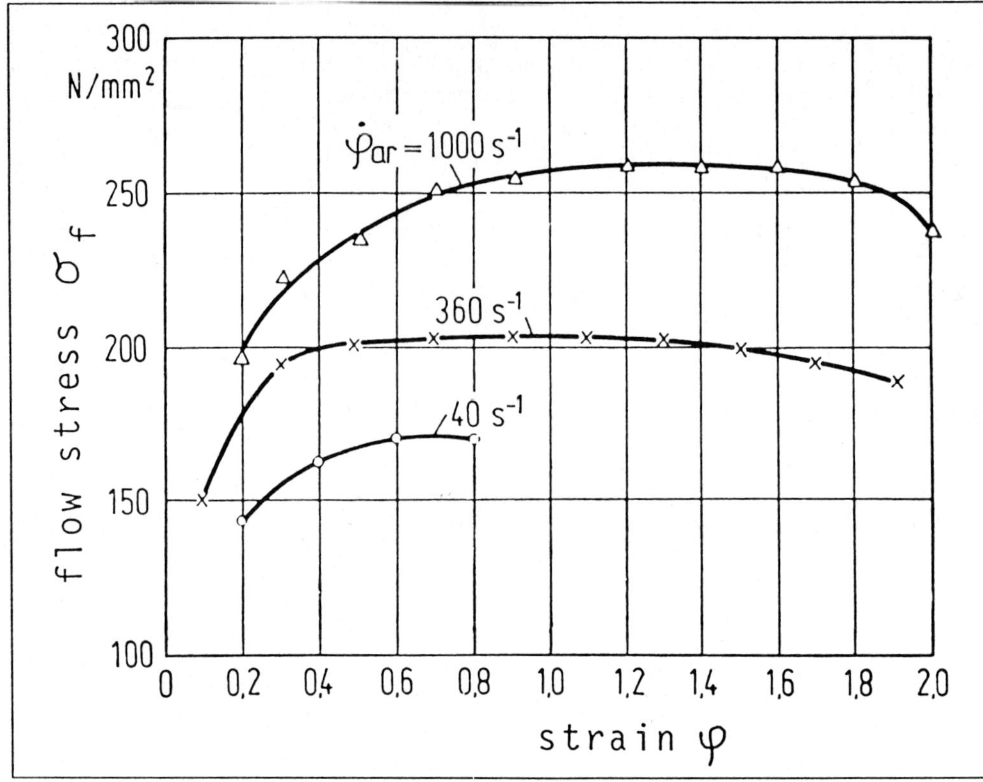

Figure 3: Effect of strain rate on flow stress of C15 steel at 1100 °C [6].

where σ_{f1} is flow stress at strain rate ϕ_1. For steels, typical values of the exponent m lie in the range -0.02 to +0.05 at 20 to 450°C (68 to 845°F), and 0.1 to 0.2 at temperatures above 880°C (1616°F). Figure 3 shows the flow curve of C 15 steel at different strain rates.

3. DETERMINATION OF FLOW CURVES

3.1 Tensile Test

In general, flow curves are determined by experiments, the most important of which are tensile test, upsetting test and torsion test [5,6,7]. A description of these experiments with special emphasis on testing at elevated temperatures is given in [9]. There is no "best" experiment since each of them has a special field of application. The proper choice of a testing method - including specimen size - depends on the metal forming process to be simulated. If the flow curve shall only be determined for small strains, the tensile test will normally be preferred because of its simplicity, and also because for this experiment, the conditions of testing have been well defined by standards. The tensile test will also be preferred if the flow curve which has been determined for low strains can be extrapolated to higher strains. For the range of uniform deformation, it is usually assumed that the stress is constant over the cross section. Hence, the flow stress is given by Equation 1 and the cross sectional area, A, by Equation 3. The range of uniform deformation is limited by that elongation for which maximum force is obtained. Beyond this elongation, deformation begins to localize by necking. For many metals equivalent strain at uniform elongation is given by the equation

$$\Phi_u \cong 0.2....0.3 \tag{9}$$

According to SIEBEL and SCHWAIGERER [10] the flow curve can also be determined for the region of necking by means of the equation (see Figure 4):

$$\sigma_f = \frac{F}{A_{min}\left(1 + \frac{a}{4\rho}\right)} \tag{10}$$

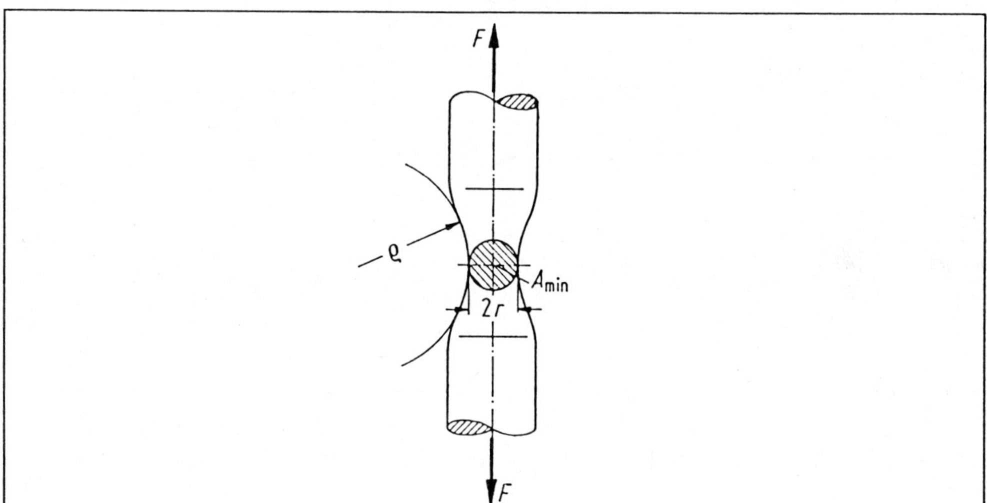

Figure 4: Necking of tensile test-piece (schematic)

The corresponding equivalent strain is given by

$$\Phi = \ln\left(\frac{A}{A_{min}}\right) \tag{11}$$

where A_{min} is the minimum cross-section in the necking zone at a given force. By this method the flow curve can be determined up to strains of the order of unity ($\phi \cong 1$).

A simplified determination of the flow curve is possible for unalloyed and low-alloyed steels at room temperature up to a strain $\phi \cong 1$. In these cases, the flow-curves fulfil the equation

$$\sigma_f(\Phi) = C\Phi^n \tag{12}$$

where C and n are specific constants of the material (n = "strain hardening coefficient"). Equation 12 shall be referred to as the LUDWIK equation [11] though it was first proposed by HOLLOMON [12]. If Equation 12 can be assumed only the constants C and n have to be determined.

3.2 Upsetting Cylindrical Test-Pieces

Normally the formability of metals is lowest at tensile hydrostatic stress [13,14]. Therefore higher strains are obtained by upsetting tests than by tensile tests. If a cylindrical test-piece is compressed between parallel dies, see Figure 5, the equivalent strain according to the TRESCA criterion is given by the equation

$$\varepsilon(F) \cong \ln\left[\frac{h(F)}{h_0}\right] \tag{13}$$

where h(F) is the specimen height at a force F. During the test, the reduction of height, i.e.

$$s(F) = h_0 - h(F) \tag{14}$$

is registered. Then

$$h(\varepsilon) \cong h_0 e^{-|\varepsilon|} \tag{15}$$

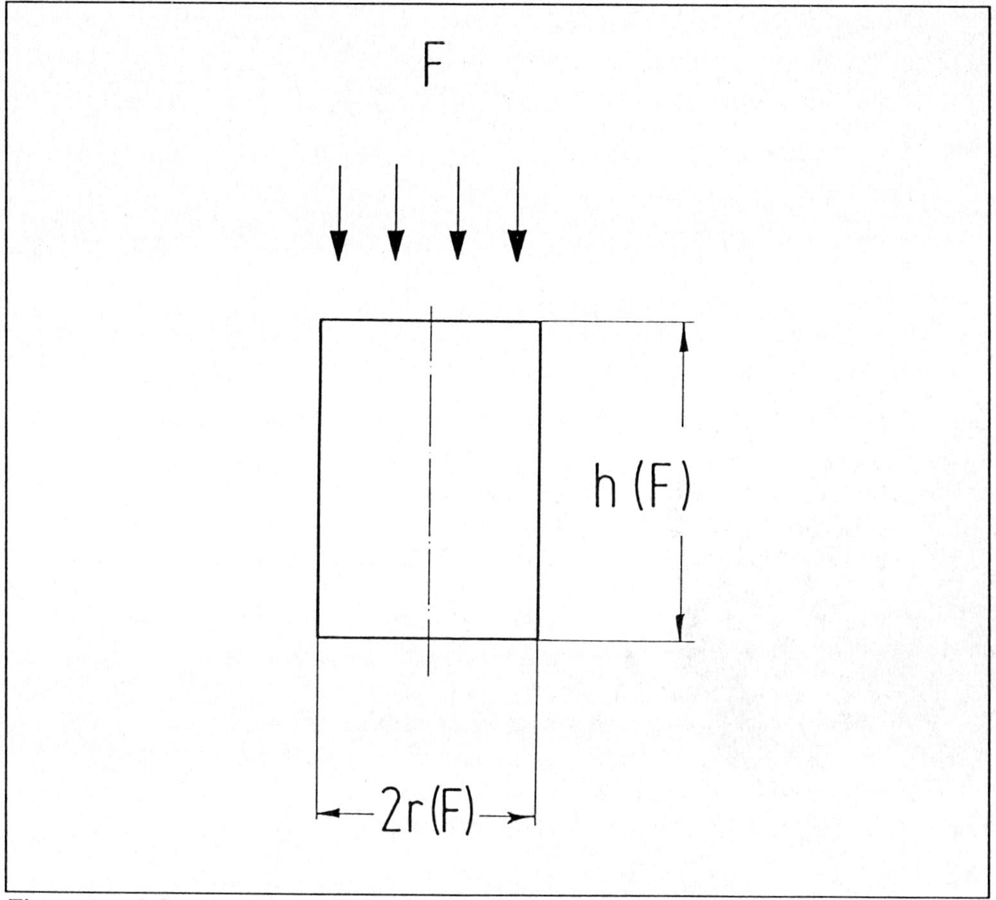

Figure 5: Scheme Of Upsetting Test On Cylindrical Test-pieces

where the equivalent strain ε is negative by definition, see Equation 13. In many cases the effect of hydrostatic stress on the flow stress can be neglected so that

$$\sigma_f(\varepsilon) = \sigma_f(-\varepsilon) \tag{16}$$

For the determination of flow stress from experimental data, it can be written as a first approximation

$$\sigma_f(F) = \frac{F(\varepsilon)}{\Pi r^2(\varepsilon)} \cong \frac{F[h_0 - s(F)]}{\Pi r_0^2 h_0} \tag{17}$$

Using Equations 13, 14 and 17, strain and stress can be calculated from the measured curve F(s). In this calculation, the error of force only propagates into stress while the error of reduction of height propagates both into stress and strain. So it can be expected that the experimental error of s is more important than that of force F.

The dominant source of error, however, is friction between the test-piece and the dies which requires an additional force for obtaining a given strain. Therefore Equation 17 has to be replaced by

$$\frac{F(\varepsilon)}{\Pi r^2(\varepsilon)} \cong \sigma_f(\varepsilon)\left[1 + \frac{2\mu r(\varepsilon)}{3h(\varepsilon)}\right] \qquad (18)$$

where μ is the coefficient of friction. The left side of Equation 18 is called resistance to deformation k_R:

$$\frac{F(\varepsilon)}{\Pi r^2(\varepsilon)} = k_R(\varepsilon) \qquad (19)$$

Due to friction the specimen does not remain cylindrical. The contour of barrelling of the specimen must be measured for correcting the test results. This causes an additional error which propagates into the calculated flow curve.

In the literature, several modifications of the upsetting test have been described by which friction is either suppressed or eliminated by the method of test evaluation [16] to [28].

The simplest way of reducing friction is to use a proper lubricant. For tests at elevated temperatures a graphite suspension (e.g. Delta 144) may be used [28].

An improved lubrication is obtained by using specimens according to RASTEGAEV [19] to [27]. In this case, the lubricant is filled into end-recesses of the test-piece (see Figure 6).RASTEGAEV specimens retain a cylindrical shape up to high strains. Unfortunately for RASTEGAEV specimens the reduction of height is measured with an increased error compared with conventional specimens since the end faces do not remain plane [19]. From this, an error of both stress and strain results which increases exponentially with growing strain. Therefore, it makes no sense to continue the RASTEGAEV test to strains higher than $|\phi| = 1.2$ to 1.5.

This error cannot be reduced by using test pieces of increased slenderness ratio since, due to perfect lubrication, the RASTEGAEV specimens would skew laterally. Therefore the condition should be fulfilled

$$h_0/2r_0 \leq 1....1.5 \tag{20}$$

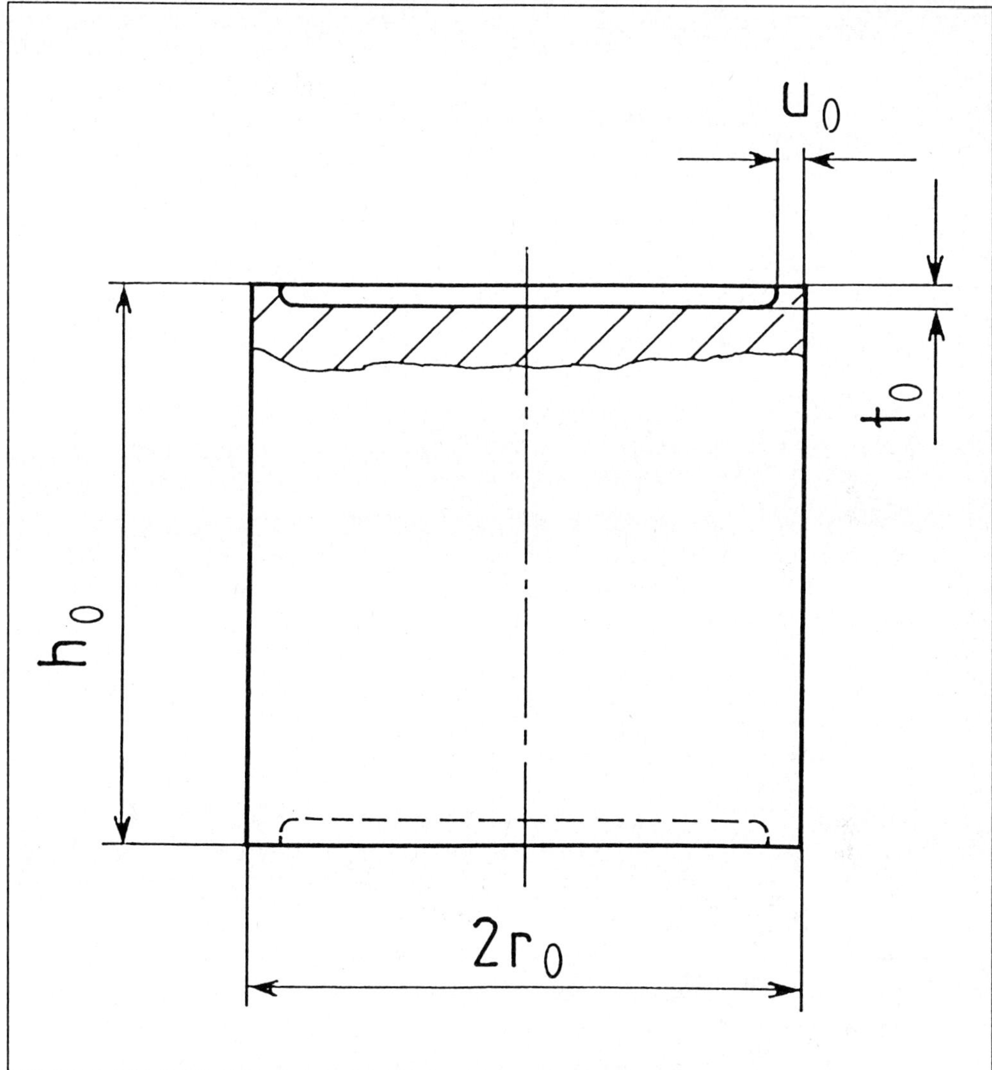

Figure 6: Upsetting test-piece according to RASTEGAEV (dimensions are, e.g.: $h_0 = d_0 = 25mm$, $u_0 = 1mm$, $t_0 = 0.4mm$).

The error can be avoided, however, by measuring the diameter instead of the height of specimen [27].

3.3 Plane Strain Upsetting Test

This experiment has been described in [29,30]. Here only the most important characteristics of the test shall be pointed out. Two adjacent rectangular dies are pressed into a strip specimen. For obtaining plane strain conditions, the ratio b/h should exceed 6. Since the area under load remains constant during the test an experimental error of the reduction of height does not propagate into stress but only into strain. The relation holds (assuming the TRESCA criterion)

$$\sigma_f(F) = \frac{F}{ab} \tag{21}$$

The dies must be kept exactly concentric since otherwise the area under load would be reduced.

The stress concentration along the edges of the dies causes initiation of fracture at a lower strain than would be the case for pure uniaxial deformation.

3.4 Effect of Strain-Rate at Elevated Temperatures

At elevated temperatures, stress is strongly strain-rate dependent. An increase of strain-rate causes an increase of flow stress, similarly to a drop of temperature [31]. Therefore the strain rate should be kept constant during deformation. However, for upsetting tests at a constant speed of movement of dies it follows from Equations 13, 14:

$$\dot{\varepsilon} = \frac{\dot{s}}{h_0} e^{|\varepsilon|} \tag{22}$$

so at a constant speed $\dot{s}$ of the die, strain rate increases exponentially with strain. For obtaining a constant strain rate, $\dot{s}$ must be reduced continuously during the test in such a way that

$$\dot{s} = \dot{s}(\varepsilon) = \dot{s}_0 e^{-|\varepsilon|} \tag{23}$$

This condition if fulfilled when using a "cam-plastometer" [31]. Since the simulation of technical hot forming processes requires a high absolute strain rate, the increase of

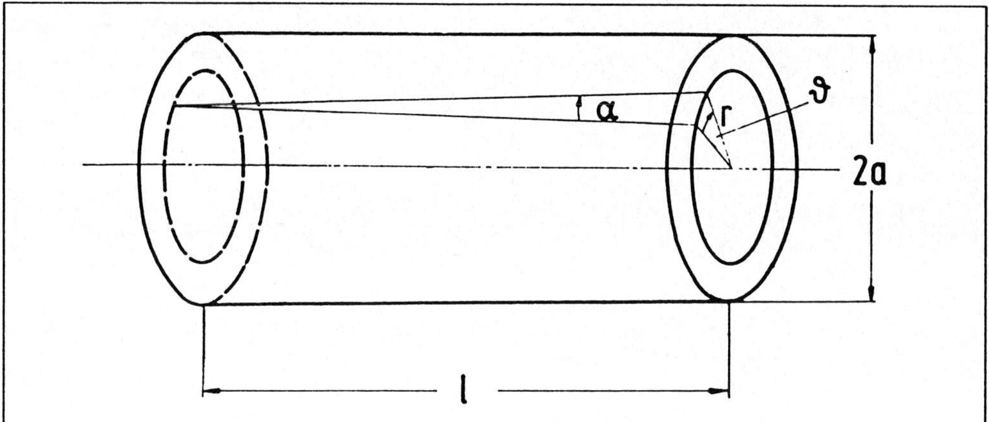

Figure 7: Scheme of Torsion Test

temperature of the specimen due to adiabatic heating must be discussed (obtaining an isothermal flow curve is only possible by interrupting the experiment at certain intervals; however, this may cause some other errors). The adiabatic increase of temperature of the test-piece has been calculated in [32].

3.5 Torsion Test

In the torsion test a cylindrical specimen is twisted by a torque acting around its axis, see Figure 7. The stress has to be calculated from the measured torque, M, and the strain from the angle of twist ψ.

Since the specimen geometry remains unchanged during deformation, the strain-rate can be kept constant more easily than in tensile or upsetting tests (by keeping the number of revolutions constant). Therefore the torsion test is especially useful for studying deformation at elevated temperatures [33]. The determination of the stress-strain curve from the test data has been described in detail in [34, 35].

Shear strain at a distance r from the axis of a long test-piece at a twisting angle is given by the equation:

$$\gamma_r(\psi) = \frac{r\psi}{l} \quad ; \qquad 0 < r \leq a \tag{24}$$

Here it has been assumed that the material is homogeneous, isotropic and incompressible, and that the length of specimen remains constant during deformation.

Conventionally strain and stress are calculated for the circumference of the test-piece (r = a). For this radial distance, Equation 24 becomes

$$\gamma_a(\psi) = \frac{a\psi}{l} \tag{25}$$

The corresponding shear strain rate is given by

$$\dot{\gamma}_a(\psi) = \frac{a\dot{\psi}}{l} \tag{26}$$

The local shear stress can be shown to be given by [34]

$$\tau(a, \gamma_a, \dot{\gamma}_a) = \frac{3M(a, \gamma_a, \dot{\gamma}_a)}{2\Pi a^3}\left\{1 + \frac{1}{3M}\left[\gamma_a \frac{\partial M}{\partial \gamma_a} + \dot{\gamma}\frac{\partial M}{\partial \dot{\gamma}_a}\right]\right\} \tag{27}$$

This calculation of local strain and stress for the surface of the test-piece is mathematically correct. However, by this, stress and strain are determined for a position where they are strongly distorted due to machining, oxidation, microcracks and notch effects which are especially strong at the surface. For these reasons, in [35,36] methods are described for calculating strain and stress at a radial distance inside the test-piece. It is shown in [35] that a "critical radius" exists for which stress is almost independent of the shape of flow curve as long as the equation is fulfilled

$$\sigma_f(\Phi, \dot{\Phi}) = \sigma_{f1}\Phi^n\dot{\Phi}^m \tag{28}$$

Figure 8 demonstrates that the curves for shear stress at different values of (n + m) - for a given torque - intersect almost exactly at one point.

The "critical radius" is given by

$$r_p = \left(\frac{3 + n + m}{4 + n + m}\right)a \tag{29}$$

Using Equation 25, the strain is obtained

$$\gamma_p = \frac{r_p}{a}\gamma_a \tag{30}$$

The corresponding shear stress can be shown to be

$$\tau_p \cong \frac{3M}{2\Pi a^3}\left(1 + \frac{n + m}{25}\right) \tag{31}$$

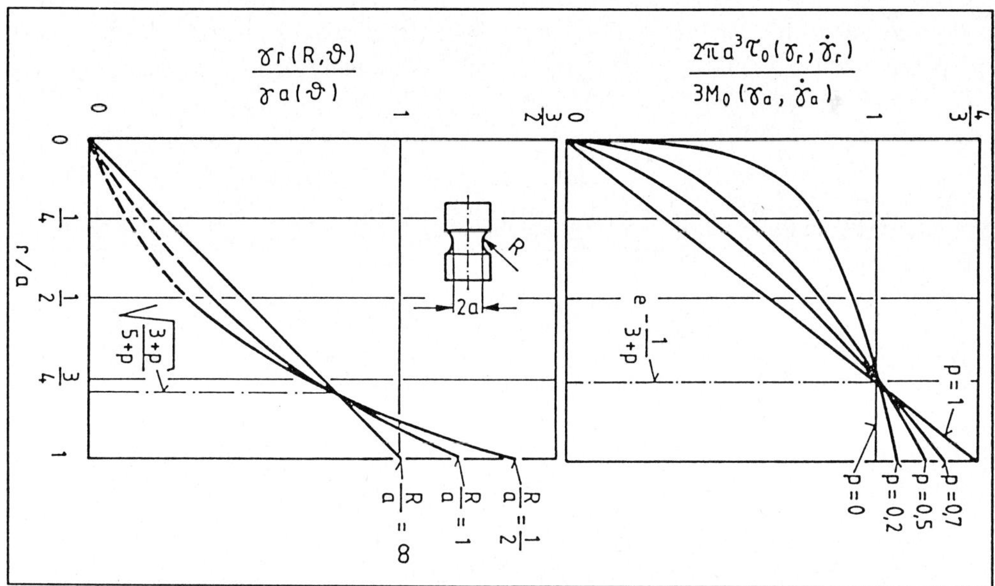

Figure 8: Shear stress in a torsion test piece for different values of $p=n+m$ *(assuming equation (28)) for the flow curve [35].*

From the curve $\tau_p(\gamma_p)$ the flow curve has to be calculated using either the v. MISES or the TRESCA criterion. In any case, equivalent strain ε is proportional to shear strain γ_p (the non-linear relation given in Ref. [31] should not be used for calculating flow curves). The difference between the v.MISES and the TRESCA criterion mainly results in a difference of absolute height of flow stress, whilst the relative course of the flow curve and strain-rate sensitivity are not strongly affected. It must be emphasized, however, that both yield criteria are based on simplifying assumptions among which that one of isotropy can cause a significant error [38]. Both criteria are also based on the assumption of strain-rate-independent flow stress. Since the torsion test is often used for just testing strain-rate-sensitive materials, this means a fundamental contradiction. However, until now, no yield criterion seems to be known which takes into account strain-rate sensitivity and which nevertheless is simple enough for practical applications.

Simulation of hot forming requires high strain rates which can be obtained by using specimens of high a/l-ratio [31], [39] to [42]. For such test-pieces the concept of "effective length" must be used for calculating strain. The "effective length" of a specimen can be determined either experimentally or semi-empirically:the notch effect which is of importance at the circumference of a short test-piece can almost be neglected when calculating stress and strain at the "critical radius" r_p [35].

4. REFERENCES

1. (Ed.) Lange, K., "Handbook of Metal Forming", McGraw Hill Book Company, New York, 1985.

2. Stüwe, H.-P., "Flow Curves of Polycristalline Metals and Their Application in the Theory of Plasticity" (in German), Z.Metallkd. 56, 1965, pp. 633-642.

3. Herbertz, R., and Wiegels, H., "The difference between Tensile and Upsetting Flow Curves as Explained by the Effect of Hydrostatic Stress" (in German), Arch. Eisenhüttenwes., 51, 1980, pp. 413-416.

4. Siebel, E. and Pomp, A., "Further Development of the Compression Test" (in German), Mitt.Kaiser - Wilhelm-Inst. Eisenforsch.10, 1928, pp. 55-62.

5. Krause, U., "Comparison of Different Methods for Determining Flow Stress in Cold Metal Forming" (in German), Dr.-Ing. thesis, Technical University, Hannover, 1962.

6. Bühler, H. and Vollmer, J., "Flow Curves of Metallic Materials at Large Strains and Strain Rates", (in German), Ind.-Anz., 91, 1969, pp. 2021-2023.

7. Rao, K.P., et. al., "Flow Curves and Deformation of Materials at Different Temperatures and Strain Rates", J. Mech. Work. Technol. 6, 1982, pp. 63-88.

8. Bauer, D., "Basic Experiments of Metal Forming" (in German), Metall, 32, 1978, pp. 778-781.

9. Ilschner, B., "High Temperature Plasticity" (in German), Springer-Verlag, Berlin/Heidelberg/New York 1973.

10. Siebel, E., and Schwaigerer, S., "Mechanics of Tensile Test" (in German). Arch. Eisenhüttenwes., 19, 1949, pp. 145-152.

11. Ludwik, P., "Elements of Technological Mechanics" (in German), Springer-Verlag, Berlin 1909.

12. Hollomon,J.H., "Tensile Deformation", Trans. AIME, 162, 1945, pp. 268-290.

13 Stenger, H., "On the Dependence of the Formability of Metallic Materials on the Stress State" (in German), Dr.-Ing. Thesis, Technical University Aachen, 1965.

14. Vater, M. and Lienhart, A., "Dependence of the Formability of Metallic Materials on the Stress State at Different Temperatures and Strain Rates", (in German), Bander Bleche Rohre, 13, 1972, pp. 387-395.

15. Pawelski, O., "Comparison of Methods for Testing the Hot Formability of Metals" (in German), in "Hot Forming and Hot Strength", Symposium Bad Nauheim 1975, Oberursel, DGM, 1976.

16. Sachs, G., Z. Metallkd., 16, 1924, p. 55.

17. Cooke, M., and Larke, C., "Resistance of Copper Alloys To Homogeneous Deformation in Compression", J. Inst. Met., 71, 1945, pp. 371-390.

18. Sato, Y., and Takaeyama, H., "An Extrapolation Method for Obtaining Stress-Strain Curves at High Rates of Strain in Uniaxial Compression", Tech. Rep. Tohoku Univ., 44, 1980, pp. 287-302.

19. Nester, W., and Pöhlandt, K., "Determination of Flow Curves in Different Modification of the Upsetting Test" (in German), Rheol. Acta, 21, 1982, pp. 409-412.

20. Suyarov, D.I., et. al., "Determination of the Flow Stress of Metals" (in Russian), Zavod lab., 21, 1955, pp. 97-99.

21. Wiegels, H. and Herbertz, R., "Upsetting Test on Cylindrical Test Pieces with High Friction for the Determination of Flow Stress" (in German), Stahl u. Eisen, 99, 1979, pp. 1380-1390.

22. Pöhlandt, K., and Nester, W., "The Determination of Stress-Strain Curves by the Compression Test, pts. I and II", Wire, 32, 1982, pp. 62-68, 150-153.

23. Rastegaev, M.V., "New Method of Homogeneous Compression of Specimens for the Determination of Flow Stress and the Coefficient of Internal Friction" (in Russian), Zavod Lab., 1940, p. 354.

24. Turno, A., "Determination of Strain-Hardening Curves Using Specimens with End Recesses" (in Polish), Obrobka Plast. (Poznan), 11, 1972, pp. 123-127.

25. Krokha, V.A., "A Method of Determining the Flow Stress in Compression to High Degrees of Plastic Deformation", Ind. Lab., 40, 1974, pp. 754-758.

26. Pöhlandt, K., "Upsetting Test for Determining Flow Curves According to Rastegaev" (in German), Ind.-Anz. 101, 1979, no. 48, pp. 28-29 (HGF 79/26).

27. Rasmussen, S.N., Pöhlandt, K., "Improving the accuracy of Upsetting Test for Determining Stress-Strain Curves", First Int. Conf. Technology of Plasticity, Tokyo, 3 -7, September 1984.

28. Diether, U., "Warm Flow Curves of Steel and their Evaluation by Compression Test" (in German), Draht,30, 1979, pp. 615-617.

29. Watts, A.B., and Ford, H., "On the Basic Yield Stress Curve for a Metal", Proc. Inst. Mech., Eng., 169, 1955, pp. 1141-1150.

30. Kaspar, R. and Pawelski, O., "A Computer-Controlled Simulation of Hot Work by Flat Compression on a High Speed Servo-Hydraulic Testing Machine", in Davies, B.J. (ed.), Proceedings of the 19th MTDR Conference, Manchester, Sept. 13.-15., 1978, London, Macmillan, 1979.

31. Sellars, C.M., Tegart, W.J., "The relation between flow stress and Structure during hot forming" (in French.), Mem. Sci. Rev. Metall. 63, 1966, pp. 731-746.

32. Lahoti, G.D., and Altan, T., "Prediction of Temperature Distributions in Axisymmetric Compression and Torsion", Trans. ASME., J. Eng. Mat. Technol. 96, 1974, pp. 1-8.

33 Weber, K.H., "The Hot Torsion Test as a Measure of Hot Formability" (in German), Thesis, Bergakademie Freiberg, 1968.

34. Stüwe, H.-P., Turck, H., "On the Determination of Flow Curves in the Torsion Test" (in German), Z. Metallkd., 55, 1964, pp. 699-703.

35. Pöhlandt, K., Tekkaya, A.E., "The Torsion Test - Plastic Deformation to High Strains and High Strain-Rates", Mater. Sci. Technol.,1, 1985, 972-977.

36. Barraclough, D.R., "Effect of Specimen Geometry on Hot Torsion Test Results for Solid and Tubular Specimens", J. Test. Eval. (JTEVA) 1, 1973, pp. 220-227.

37. Nadai, A., "Theory of Flow and Fracture of Solids", New York, McGraw Hill, 1950.

38. Witzel, W. "Inhomogeneous Deformation Due to Changes of Texture" (in German), Z. Metallkde., 69, 1978, pp. 337-343.

39. Pöhlandt, K., "Testing Strain-Rate Sensitive Materials in the Torsion Test", Materialprüfung, 22, 1980, pp. 399-406.

40. Itihara, M., Rep. Tohoku Univ., 11, 1935, pp. 489-528, cited in Siebel, E., "Handbook of Materials Testing" (in German), vol. 2, Springer-Verlag, Berlin, 1939.

41. Weiss, H., et. al., "A Torsion Machine for Programmed Simulation of Hot Working", J. Phys. E. Sci. Instrum., 6, 1973, pp. 710-714.

42. Pöhlandt, K., "On the Optimization of Specimen Shape and the Evaluation of Data from the Torsion Test" (in German), Dr. -Ing. Thesis, Technical University Braunschweig, 1977.

Chapter 3

Metallurgical Study of the Hot Upsetting of 1035 Steel

by R.A.Cohen, submitted by B.F.von Turkovich

1. ABSTRACT

The metallurgical behavior of AISI 1035 steel subjected to axially symmetrical upsetting at high temperatures was studied. Attention was paid to the existence of dynamic recovery and recrystallization. The resulting microstructure was found to depend, in a significant manner, on strain rate and, of course, on temperature.

2. INTRODUCTION

Hot-working usually takes place at temperatures of .6 T_m, where T_m is the melting temperature in K (JONAS, SELLARS and TEGART [1]). Conditions may exist for the onset of recovery and recrystallization while deformation is taking place. In 1968 STÜWE [2] questioned whether new grains observed after hot-working were formed during or after deformation. In the same volume JONAS et al. [3] discussed recovery during hot-working. Investigators began looking for evidence of dynamic recovery and recrystallization which occur during deformation as distinguished from static recovery and recrystallization discussed above.

Dynamic recovery, like static recovery, involves dislocation cross glide, climb and annihilation. As a result, strain hardening during deformation is reduced. The cold-worked cell structure gradually polygonizes to subgrains as the metal softens (MCQUEEN [4]). The original grain structure is preserved. JONAS [1] described evidence of dynamic recovery in the case of ferrous materials by using etches capable of attacking low-angle boundaries to visualize the subgrains. In aluminum, electrolytic etching has been used. Both methods can be sensitive to impurities and alloy content. X-ray microbeam and electron microscope techniques have been utilized with more reliable analysis of subgrain size and orientation.

Dynamic recovery principally occurs in pure fcc and bcc metals where medium to high stacking fault energy limits the extension of the dislocations and facilitates thermal activation of cross-glide and climb.

In 1968 MCQUEEN [4] explored the question of whether recrystallization occurs during hot-working. In his investigation of hot-rolled fcc metals with a range of stacking fault energies MCQUEEN made the following conclusions:

1. With rapid quenching after hot-rolling, the fraction recrystallized increased with decreasing stacking fault energy.
2. Worked regions showed increased polygonization with increased stacking fault energy and temperature.
3. Recrystallization occurred after the deformation ceased.
4. Dynamic recrystallization should be inhibited by high strain rate.
5. Dynamic recovery is the mechanism in fcc hot-worked metals up to a strain of 2 and probably beyond.

In 1968 MCQUEEN [4] again stated that recrystallization is not expected during hot-working. The reason offered being the retarding effect of simultaneous deformation on the nucleation of new grains. MCQUEEN continued, saying that a nucleus forms by the coalescence of several neighboring subgrains with the released dislocations forming the outer boundary which increases in misorientation to the extent it can migrate. Deformation, however, redistributes the dislocations to maintain all boundaries statistically similar in misorientation therefore inhibiting the coalescence.

In 1969 JONAS et al. [1] discussed the evidence available concerning the likelihood of dynamic recrystallization. JONAS and his coworkers described two broad groups of metals and alloys. The first group develops subgrains when cooled rapidly after either small or large deformations. The second group shows subgrains after low strains but after higher deformations they develop an equiaxed recrystallized grain structure. This structure could result from either static or dynamic recrystallization, although JONAS et al. offered the following convincing evidence in support of dynamic recrystallization:

1. Upon examination of a recrystallized structure after high strains, some of the annealing twins were distorted suggesting that deformation occurred after the grains recrystallized.
2. A substructure was observed within the recrystallized grains which is in contrast to the dislocation-free grains visible after static recrystallization.
3. There were similarities between the regular cycles in flow stress and the cycles in creep rate associated with dynamic recrystallization.

In 1979 MCQUEEN [5] concluded that dynamic recrystallization occurs in some alloys in the course of straining when the dislocation density becomes high enough to nucleate new grains. In discussing the use of TEM in the metallography of hot-working he stated that the freshly nucleated dynamically recrystallized grains have a recovered substructure not found in statically recrystallized grains. The substructure is not uniform from grain to grain as it increases with the time or strain which has passed since the nucleation of the grain; the freshly nucleated grains being recognized by size and low density of substructures. He concluded that modern metallographic techniques along with stress-strain curves demonstrate the existence of dynamic recovery and recrystallization during hot-working.

In 1979 MCQUEEN and BAUDELET [6] compared the microstructures in dynamic recovery and recrystallization. Table I, based on their tables, summarizes the prerequisite, steady state, and final microstructures observed under the two conditions.

The occurrence of dynamic recrystallization has become more accepted. In 1984, SAKAI and JONAS [7] published a useful summary. In materials of high stacking fault energy, recovery processes completely balance the effects of work hardening. In materials of moderate to low stacking fault energy, the dislocation density increases to appreciably higher levels, eventually permitting dynamic recrystallization.

It must be kept in mind that strain, strain rate and temperature can vary in different regions of a hot-worked specimen. POKORNY [8] described the recrystallization of the more highly worked areas in a forging during reheat to illustrate the differences which may exist. In the 1035 specimens in this study, tested under a moderate true strain of 0.47, the shear areas near the dead zones, the central areas and the outer radius are worked most to least respectively. Temperature differences also occur with the central region maintaining temperature more effectively. The local conditions of strain, strain rate and temperature in the hot-worked specimen determine the final microstructure which may not be uniform throughout the specimen.

It is understandable that interpretation of the microstructure of a hot-worked material can be difficult. Not only is there deformation taking place but simultaneous recovery and/or recrystallization, depending on the working history, the material and the cooling rate may occur. Upon cooling some materials will also be undergoing transformations such as fcc austenite to bcc ferrite and pearlite in steels.

Yet understanding the hot compression test microstructure is necessary for assessment of industrial hot-working processes in order to obtain optimum commercial results as well as to expand basic knowledge of material behavior.

3. PLAIN CARBON STEEL

3.1 General Properties

AISI 1035 is a highly versatile, medium carbon steel. The chemical composition range for 1035 is: C .32-.38%, Mn .6-.9%, P_{max} .04%, S_{max} .05% (HOYT [9]). It has moderate strength and hardness in the as-rolled or as-forged condition and can be strengthened by cold-drawing or heat treatment. 1035 steel is often used for forging or for parts machined from bar stock. Typical uses for 1035 are shafts, bolts and other medium-strength parts. When quenched it is used for gears and sprockets of moderate strength. It is available commercially in hot rolled bars, billets, slabs, plate, sheet, strip, wire, cold rolled bars, cold drawn bars as well as shapes (CARMICHAEL [10]).

The specimens in this study were flame cut from 1035 plate. In platemaking, the ingots are normally passed through a slabbing mill where the thickness is reduced. Sometimes the plate is hot scarfed to remove surface imperfections. Then the plate is reheated and rolled to the final thickness.

An alternative method is to continuous cast the steel, which is then heated and rolled to the final thickness in one operation, then roller leveled and cooled.

Depending on the thickness, the plate is sheared or flame cut to size. If no further processing such as heat treatment is needed the plate is tested and shipped.

Steel plate can exhibit anisotropic properties called directionality or fibering. Inclusions which are plastic at rolling temperatures are elongated in the direction of rolling. Localized chemical segregates formed during solidification of the steel are also elongated. Although this directionality reduces ductility and impact properties transverse to the rolling direction, there is little or no effect on strength (BARDES [11]).

Confining the discussion to 1035 steel under equilibrium conditions the important phases to consider are ferrite, austenite and pearlite.

We can follow the changes occurring upon heating a 1035 steel under equilibrium conditions by following the vertical line on the phase equilibrium diagram. Starting with ferrite and pearlite, at 600°C the ferrite holds about .007 w/o C. Up to 727°C the solubility of carbon in ferrite increases until it can hold about .025 w/o C. The first phase change occurs at 727°C which is called the A_1 critical temperature above which all the pearlite transforms to austenite. The transformation begins with small grains of austenite formed within the pearlite regions. The boundaries between the carbides and ferrite in pearlite are regions of high energy. Upon heating, with the change of bcc ferrite to fcc austenite, diffusion of carbon into the fcc lattice occurs, lowering the energy. As the temperature is raised, the austenite grains grow until all the pearlite is transformed and the austenite encroaches into the ferrite grains (TEICHERT [12]). As the temperature rises through the dual phase region, the austenite dissolves more and more of the proeutectoid ferrite until at A_3 the last of the proeutectoid ferrite has been absorbed into the austenite which will have the same average carbon content as the steel.

Upon slow cooling under equilibrium conditions the reverse occurs. At approximately 815°C (A_3 for 1035 steel), the austenite starts to reject free ferrite, usually at the grain boundaries (SAMUELS [13]). The fcc iron changes to bcc. Below A_1 the austenite becomes progressively richer in carbon until just below the eutectoid temperature, A. When the remaining austenite reaches the eutectoid composition of .8 w/o C the simultaneous rejection of both ferrite and carbide will occur producing pearlite. In summary, the decomposition of austenite involves the allotropic transformation of iron and the rejection of carbide at the eutectoid (TEICHERT [12]). The final structure is a mixture of ferrite and pearlite (MCGANNON [14], BAUMEISTER [15]). In the case of 1035 steel, there will be approximately 41% pearlite and 49% ferrite using the lever rule and the phase diagram.

Although the new structure has its own specific grain size and shape properties, some of the dimensional characteristics of austenite may be inherited by the new structure. The rejection of ferrite usually starts at nuclei in the austenite grain boundaries, so although a pure grain effect is only seen in eutectoid steels, there can still be a relationship between the original geometry of the austenite grains and the final ferrite and pearlite structure (SAMUELS [13]). The rate of cooling and the rejected grain boundary ferrite will also influence the relationship, with the amount of free ferrite rejected during cooling dependent upon the % carbon of the steel.

The above discussion concerned a slowly cooled 1035 steel under equilibrium conditions. Under nonequilibrium conditions, time plays an important role in the transformation because it introduces a time constraint on diffusion. If the cooling is quite rapid, only part of the ferrite is ejected to the boundary due to the limitations on the diffusion of carbides. This decreases the volume of free ferrite, while the austenite contains excess ferrite and will form a finer lamellar pearlite (SHEWMON [16]).

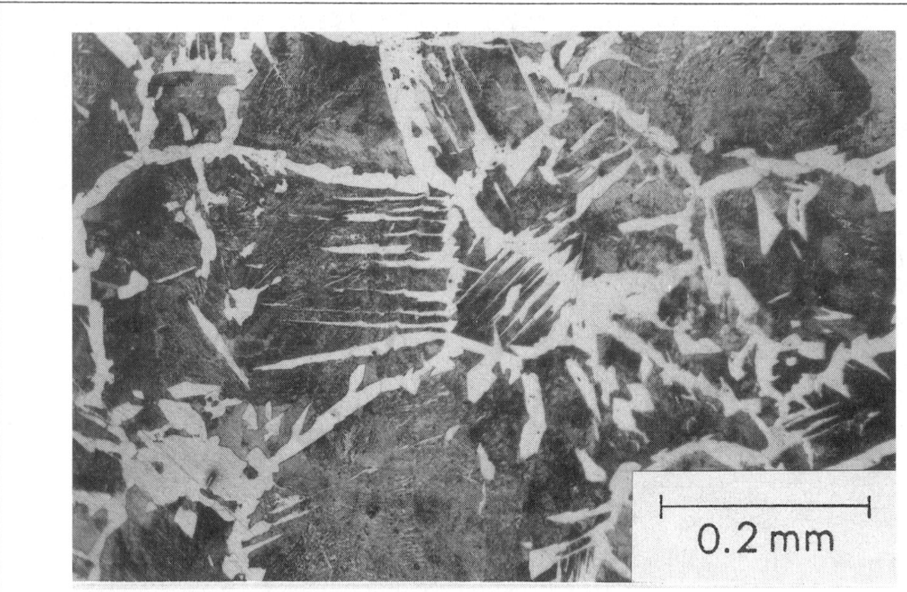

Figure 1: Microstructure of slowly cooled 1035 steel.

Slower cooling will result in more complete rejection of ferrite which will be located either at the austenite grain boundaries or be intracrystalline as in the Widmanstatten pattern.

We can describe the transformations more realistically using either an isothermal transformation diagram or a continuous cooling transformation diagram (CCT).

Because under most industrial conditions the reaction does not go to completion under isothermal conditions, the CCT diagram is more realistic to use for describing cooling processes.

The hot-worked air cooled 1035 specimens in this study contained ferrite and pearlite as shown in Figure 1.

3.2 Effects of Cold Work and Subsequent Recovery or Recrystallization on Microstructure

Since the cold-working produces strain hardening, the density of dislocations increases and an intragranular cell structure forms. The average spacing between cell walls decreases as strain increases. The movement of the dislocations is determined by crystal structure and the mode of deformation such as slip or twinning. MICHALAK [17] described the crystallographic planes in fcc and bcc crystals on which the dislocations move. Slip occurs in the bcc lattice of iron or ferritic steels by glide of dislocations with Burgers vector (unit displacement) of type $a/2 \langle 111 \rangle$ on $\{110\}$, $\{211\}$ and $\{321\}$ planes giving a total of 48 possible slip systems. Slip may occur with equal ease on any plane containing a $\langle 111 \rangle$ slip direction. In the fcc lattice of austenitic steels, slip occurs by movement of dislocations with Burgers vector of type $a/2 \langle 110 \rangle$ on $\{111\}$ planes giving a total of 12 possible slip systems.

Steel in the cold-worked state will have a preferred crystallographic orientation or texture. During deformation the lattice rotates toward one or more stable orientations thus developing a texture. The development of texture depends on (LESLIE [18]):

1. Type of deformation
2. Amount of deformation
3. Temperature of deformation
4. Mechanics of slip in the metal i.e. fcc, bcc, etc.

Figure 2 illustrates the time course of recovery for iron. The rates of change decrease sharply initially, become more gradual and approach zero slowly. The initial recovery becomes more rapid:

1. with an increase in temperature
2. with greater amounts of stored energy i.e. prior cold-work
3. with diminishing grain size because the dislocation density is greater for a given strain.

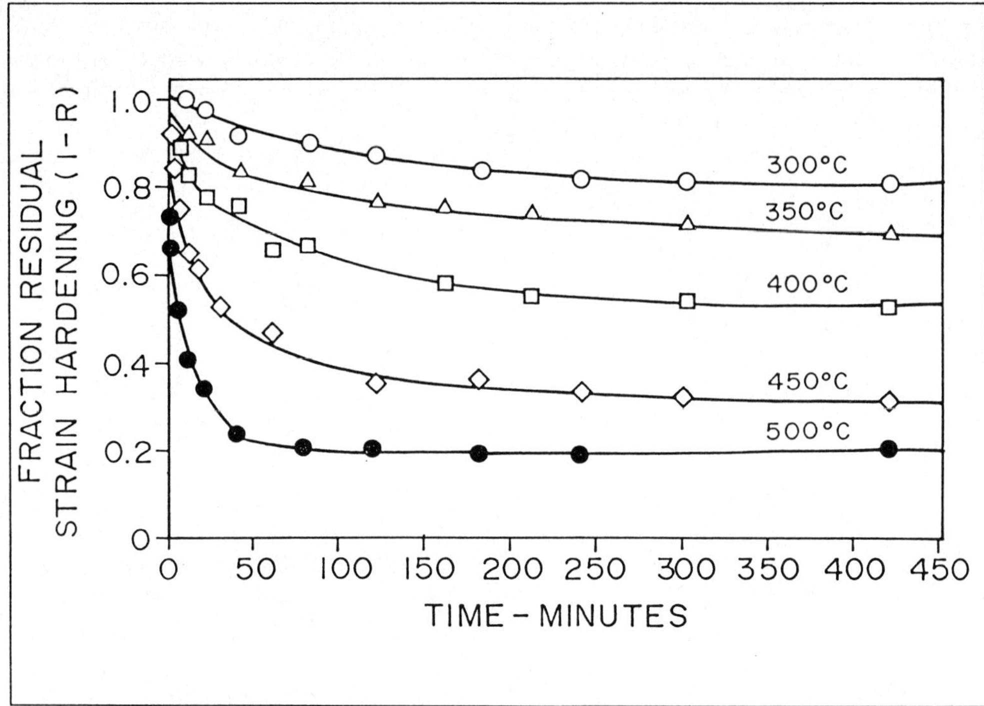

Figure 2: Recovery of iron after 5% prestrain.

During recovery, which takes place below the recrystallization temperature, no major changes occur in the texture developed during cold-working (JOLLEY and WITMER [19]). A recovered microstructure retains the original cold-worked elongated grain boundaries but forms a substructure of cells or subgrains. In contrast, a recrystallized microstructure is identified by equiaxed grains with low dislocation density. A worked grain provides more sites for nucleation. Nucleation can occur intragranularly as well as along the austenite grain boundaries.

3.3 Microstructures Developed During Hot-Working

The above discussion was concerned with static heating after cold-working of 1035 steel. When 1035 is hot-worked, dynamic processes must be considered. During deformation, strain hardening is taking place along with dynamic recovery and recrystallization. After deformation, static recovery and recrystallization may occur. During cooling, a phase transformation from fcc to bcc occurs.

LESLIE [18] offered a classification of thermomechanical treatments of steel summarized in Table II. The microstructure of interest developed in hot compression resulted from deformation in the supercritical region as the specimens in this study were tested in the austenite state.

Table III, based on TEGART and GITTINS [20], summarizes the restoration processes associated with hot-working of austenite.

As discussed above, during transformation the austenite grain boundaries provide sites for nucleating ferrite thus the final austenite grain structure directly influences the final room temperature structure. TEGART and GITTINS [20] propose three possible structures for the austenite after hot-rolling above the transformation temperature.

1. Fully recrystallized equiaxed structure.

   At higher temperature and slow cooling rates to the transformation temperature, the austenite grain's size may increase before transformation begins, resulting in larger ferrite grains. Since transformation is occurring in austenite with a relatively low dislocation density, the ferrite grain size depends on the austenite grain size and the rate of cooling through the transformation range.

2. Unrecrystallized, elongated structure.

   The final structure will be fine grained and equiaxed because there are more ferrite nucleation sites.The high dislocation density and the large austenite grain boundary area result in rapid nucleation and growth of ferrite. The final grain size will also depend on the amount of deformation below the recrystallization temperature and on the rate of cooling through the transformation range.

3. Partially recrystallized structure.

   The recrystallized austenite may be different in size from the unrecrystallized austenite. This will transform into a mixed size ferrite grain structure having poor properties.

As the fully recrystallized specimen is slowly cooled from the temperature of deformation, there will be some grain growth. As the transformation temperature is reached, the ferrite and pearlite grain size will be related to the original equiaxed austenite grain size. At lower temperatures of deformation there is less time for austenite grain growth and in general the smaller austenite grains result in a finer structure after transformation (TEICHERT [12]).

As JONAS et al. [1] discussed, the recrystallization kinetics during rolling is controlled by precipitation of carbides or carbonitrides in the austenite. Reheating temperature determines the solution rates of carbides. The amount of deformation per rolling pass affects the amount and mode of precipitation of carbides. Finally, the finish rolling temperature alters the amount of precipitation on further cooling.

In addition to the TEGART and GITTINS description above, an unrecrystallized, elongated austenite can take on a copper texture {124}<112> to {146}<112> (LESLIE [18]). As discussed above, the final grain structure will be fine grained and equiaxed.

In partially recrystallized austenite, there will be different amounts of deformation in the austenite grains depending on when and if they are recrystallized grains. This results in a difference in dislocation densities and nucleation sites causing mixed grain sizes in the final structure.

In 1979, BROWN and DEARDO [21] studied the behavior of austenite during hot-rolling to clarify the mechanism leading to the observed equiaxed microstructure. Using optical and transmission electron microscopy as well as x-ray pole figures, they concluded that dynamic recrystallization is the responsible mechanism but the conditions for dynamic recrystallization in steels are still not clearly defined. Table IV presents some examples of observations of dynamic recrystallization in steels.

In 1985, BERNSHTEIN et al. [22] observed only dynamic recovery in the austenitic stainless steel they tested. The testing conditions were: strain .07-.1, strain rates .001 sec^{-1}, .1 sec^{-1}, and 1.0 sec^{-1} and temperature of 1050°C for compression and 1025°C for rolling. BERNSHTEIN et al. [22] acknowledged that many observations of dynamic recrystallization have been made. They concluded: the austenitic steels are, apparently, capable of softening by either mechanism, depending on their composition and the conditions of hot deformation.

The final structure and properties of the hot-worked steel depend upon the static and dynamic restoration processes occurring in the hot-worked austenite microstructure. We must control the microstructure going into the transformation in order to control the final microstructure and properties coming out of the transforma-

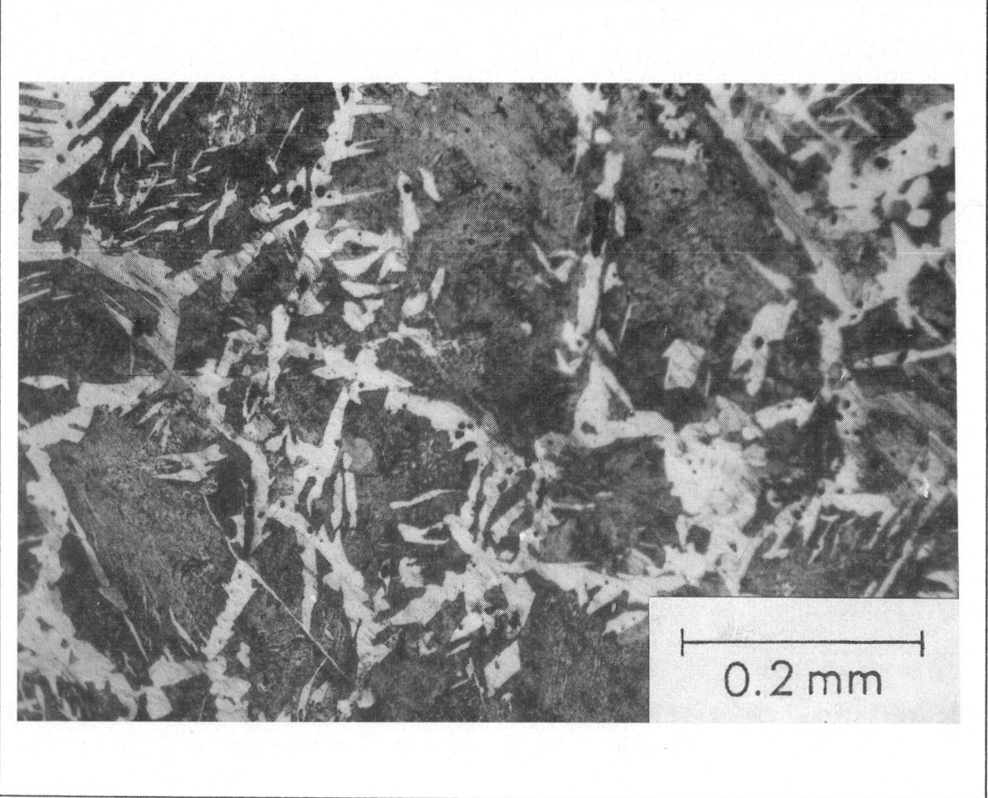

Figure 3: Original material as cut from plate.

tion. By controlling the hot-working operations, the microstructure and therefore the properties of the steel can be controlled. The result is improved combinations of high strength and toughness unavailable by just varying chemical composition.

4. SPECIMEN PREPARATION AND TESTING

In order to examine the microstructure of 1035 steel under hot-working conditions, specimens were cut from AISI plate (by flame cutting) normal and parallel to the rolling direction and were then turned on a lathe until the final dimensions of h_o = 38.1 mm (1.5") and d_o = 12.7 mm (.5") were reached. Micrographs of the initial condition of the material are provided in Figure 3. The microstructure is similar in sections which are parallel and normal to the rolling direction, showing large equiaxed grains. The cleavage or Widmanstatten pattern indicates slow cooling (TEICHERT [12]).

This structure results from intracrystalline precipitation of the ferrite along certain crystallographic planes.

Compression tests were conducted on a closed loop control MTS machine with tungsten carbide plates and graphite lubricant. (The compression tests were conducted by Prof. Kaftanoglu and his students.) Exploratory tests at constant strain rate were performed using a specially designed circuit. The results were compared with constant velocity tests and no significant differences were observed. The remaining tests were conducted under constant velocity i.e. average strain rate conditions. The average strain rates were .5 sec^{-1}, 1.0 sec^{-1}, 1.5 sec^{-1}, 3.0 sec^{-1}. The specimens were compressed from h_o = 38.1 mm (1.5") to h = 23.8 mm (15/16") for a true strain of .47 compression.

The specimens, insulated with ceramic wool, were heated to the test temperature and held at that temperature for a minimum of one hour. The platens were heated and the tests were conducted at room temperature. The insulated specimens were then allowed to air cool. Test temperatures were 816°C, 871°C, 927°C and 982°C which correspond to homologous temperatures $0.6T_m$, $0.64T_m$, $0.67T_m$ and $0.7T_m$ respectively. Table V is a summary of strain rates, temperatures, and specimen orientation used in this study.

After testing, the cylindrical specimens were cut longitudinally and the flat surface was prepared for light microscopy by mechanical polishing and a 2% Nital etch. Micrographs of the central region of each specimen were used for comparison.

To examine the effects of static heating, insulated specimens of the original material were held at the test temperatures for 90 minutes. After air cooling they were prepared for light microscopy as described above.

The material had a similar appearance in both the longitudinal and transverse orientations to the rolling direction. This equiaxed microstructure suggests recrystallization during hot-working. The large pearlite masses framed by ferrite suggest large equiaxed austenite grains before transformation.

When the original material was statically heat treated at the test temperatures, the final grain size increased with temperature as seen in Figure 4a-d. At the lowest temperature, 816°C, there are very small pearlite and ferrite areas with no evidence of the original large grains seen in the specimens before heating.

At 871°C, the grains are larger, suggesting larger austenite grains before trans-

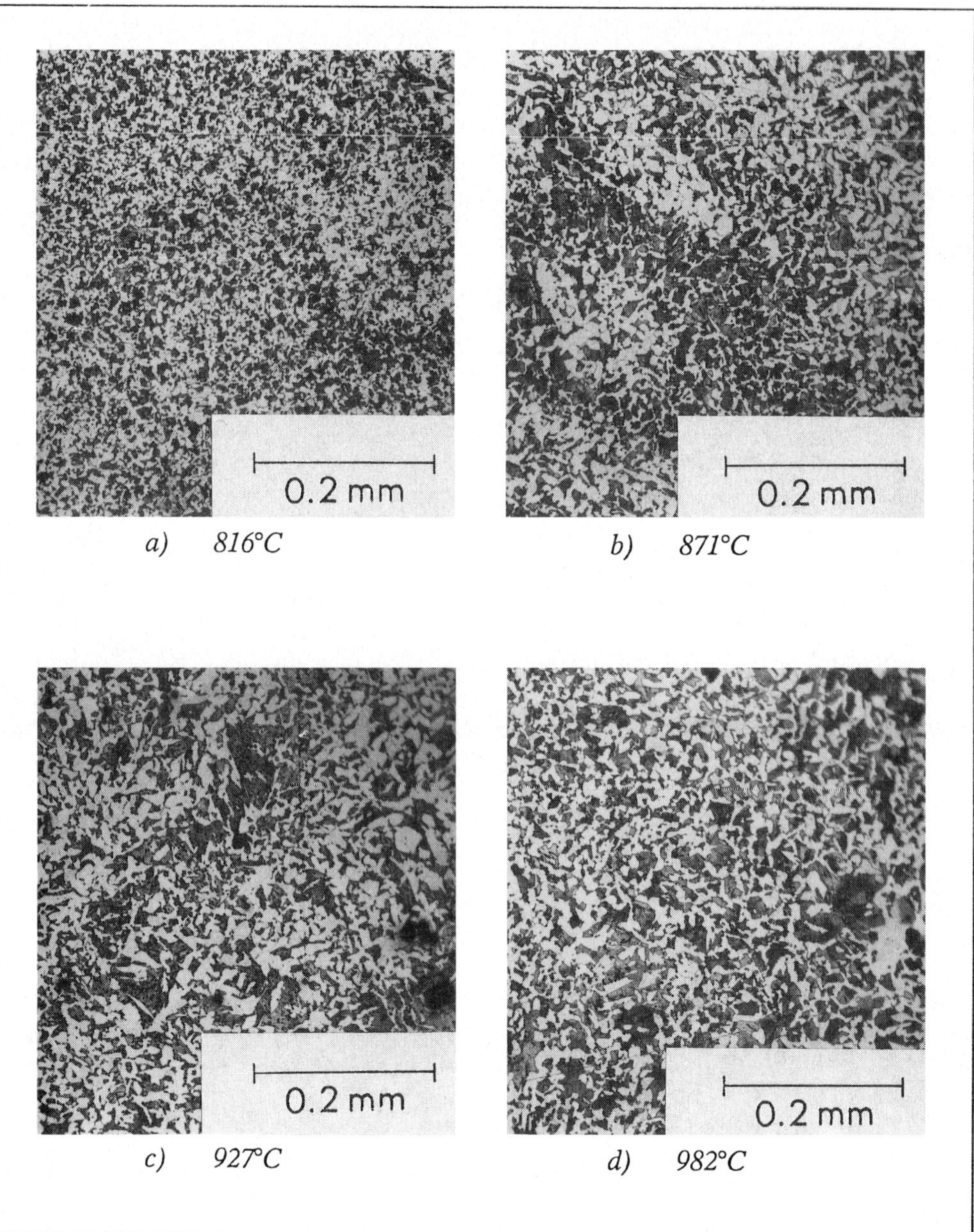

Figure 4 a-d: Statically heated specimens.

formation. There are areas richer in ferrite, perhaps left from the directionality in the original material which was not apparent with light microscopy inspection.

At 927°C, the microstructure looks similar to 871°C but with some isolated large pearlite masses near stringers. Elsewhere, the microstructure is more uniform than at 871°C, with some variation in pearlite size but no evidence of banding.

At the highest temperature, 982°C, there are some large pearlite areas scattered among the uniform areas of ferrite and pearlite.

Microscopically, the appearance of the hot compression test specimens varied. In most specimens, barrelling was not uniform. Some specimens showed a classic cone-shaped dead zone while others did not. These variations may be due to differences in lubrication, platen temperatures, asymmetry of loading or material history. Even though all specimens were obtained from the same piece of plate, there can be variations in structure within a plate depending on the history and rolling conditions.

Micrographs of the central regions of the hot-worked specimens demonstrate the trends with variations in temperature and strain rate. Figure 5a shows the resultant microstructure at the highest temperature, 982°C, and strain rate, 3.0 sec^{-1}. Medium sized equiaxed pearlite masses and ferrite grains suggest transformation from a relatively uniform austenite phase. There are also areas of larger pearlite masses in the central regions adjacent to the dead zones. These large pearlite masses are evidence of austenite, recrystallized. Figure 5b shows a similar microstructure in a specimen worked under the same conditions but with longitudinal orientation to the rolling direction.

Staying at the highest temperature of 982°C but reducing the strain rate to 1.0 sec^{-1} and .5 sec^{-1} resulted in the microstructure in Figures 5c and 5d respectively. The microstructure at a strain rate of 1.0 sec^{-1}, looks similar to that at the highest strain rate of 3.0 sec^{-1}, with areas of large pearlite masses among the medium sized pearlite and ferrite. Lowering the strain rate to .5 sec^{-1} caused a marked change in the microstructure. The pearlite masses remain medium sized but are rounder with only narrow areas of ferrite between them.

Reducing the temperature of hot-working to 927°C resulted in the microstructures of Figure 6a-6d. At a strain rate of 1.5 sec^{-1}, the average grain size is reduced with a uniform appearance throughout the specimen. At the lower strain rate of 1.0 sec^{-1}, the microstructure is similar but with some evidence of directionality especially in the

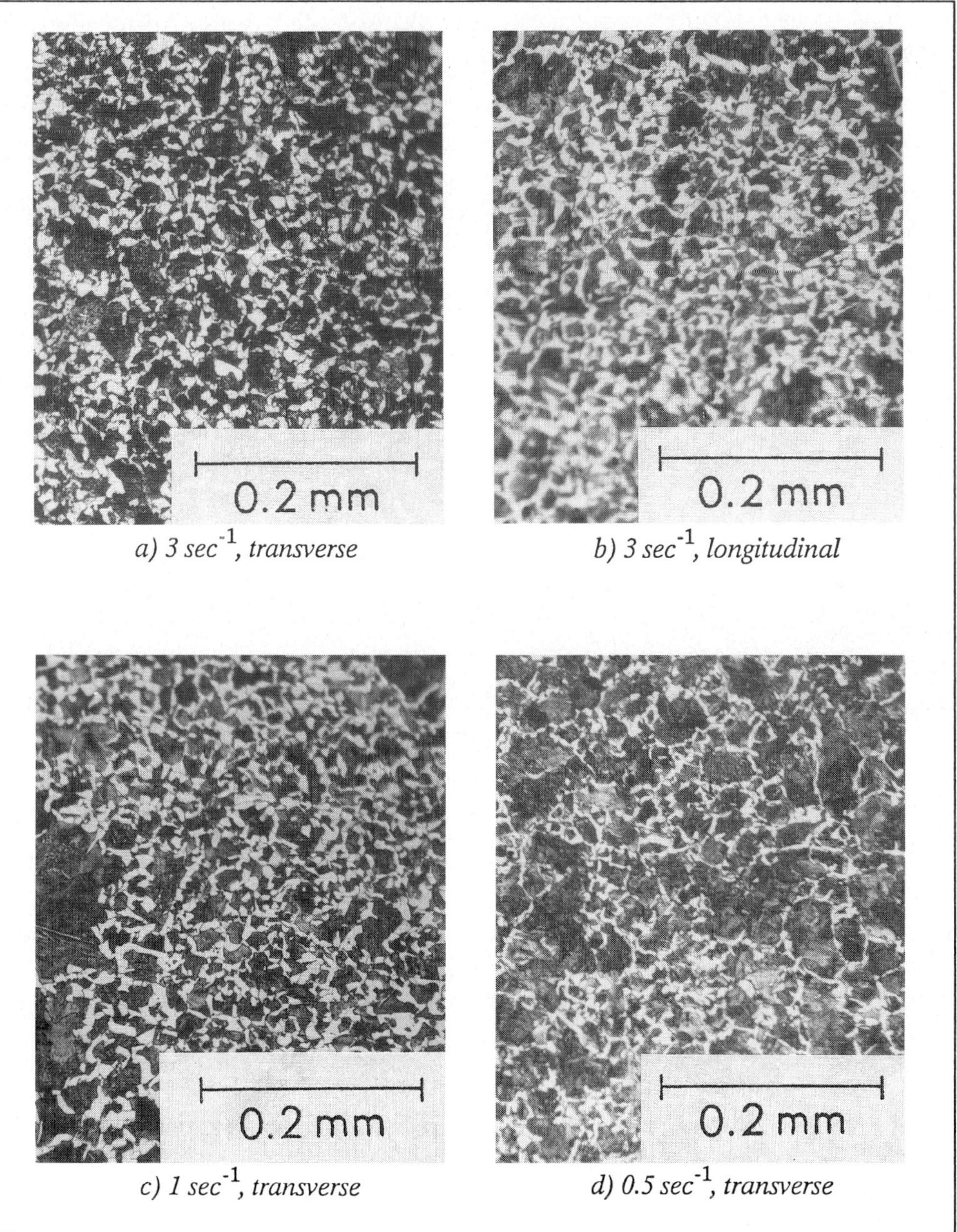

a) 3 sec^{-1}, transverse

b) 3 sec^{-1}, longitudinal

c) 1 sec^{-1}, transverse

d) 0.5 sec^{-1}, transverse

Figure 5 a-d: Compression specimens, 982°C.

longitudinal view. Finally, at the lowest strain rate of .5 sec^{-1}, the microstructure is again similar and shows some directionality.

At the lower temperature of 871°C, in Figure 7a-7d, the grains are smaller than at 927°C and 982°C. There is no major difference in microstructure among the different strain rates at this temperature.

The microstructure showed some changes at the lowest temperature of hot-working, 816°C, (Figure 8a-8d). At the two higher strain rates of 1.5 sec^{-1} and 1 sec^{-1} the microstructure is similar to those described above but with a slightly smaller average grain size. At the lowest strain rate of .5 sec^{-1} and temperature of 816°C, the microstructure shows elongated ferrite and pearlite, evidence of deformation taking place after transformation.

In summary, the hot-worked 1035 steel showed temperature effects on grain size and strain rate effects on grain size distribution at the highest temperature tested.

The statically heated specimens were compared with the corresponding dead zones as shown in the micrographs of Figures 9-10. Since the dead zones are areas which are not deformed during the compression test, they should show similar microstructures to the statically heated specimens.

The amount of time held at the test temperatures as well as the cooling rates may differ because the tests were done at different times and the exact timing of the compression tests were not available to the author. The differences in timing may account for the small differences in microstructures.

The microstructures of the constant strain rate and constant velocity (average strain rate) specimens were compared. Although the original investigators reported no differences in the behavior between the two groups, the microstructures show differences as seen in Figure 11.

In summary, the major trends observed in the microstructures included:

1. Strain rate effects on grain size distribution at high temperature.
2. Temperature effects on grain size.
3. Transformation during deformation at low temperature and strain rate.

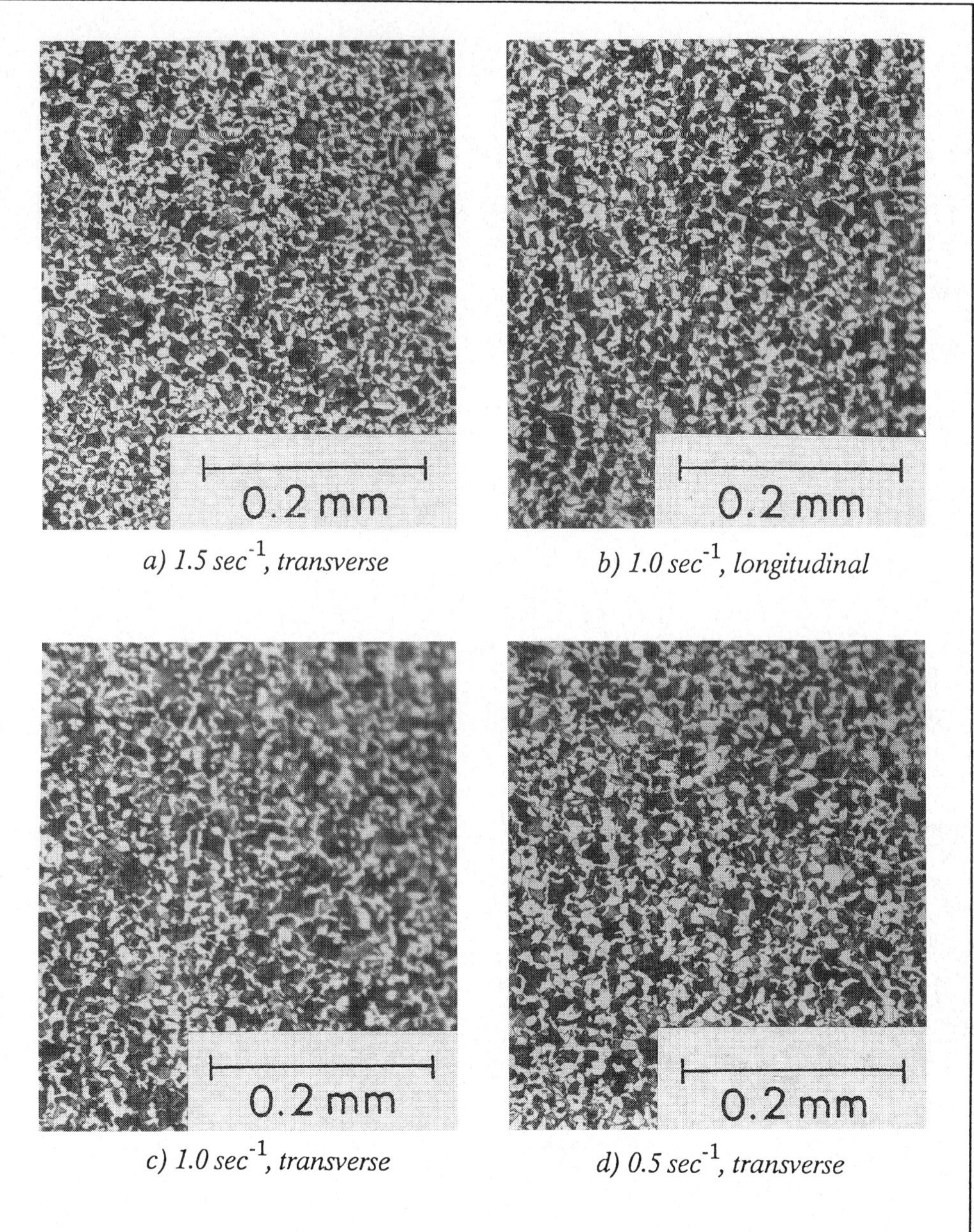

a) 1.5 sec^{-1}, transverse

b) 1.0 sec^{-1}, longitudinal

c) 1.0 sec^{-1}, transverse

d) 0.5 sec^{-1}, transverse

Figure 6 a-d: Compression specimens, 927°C.

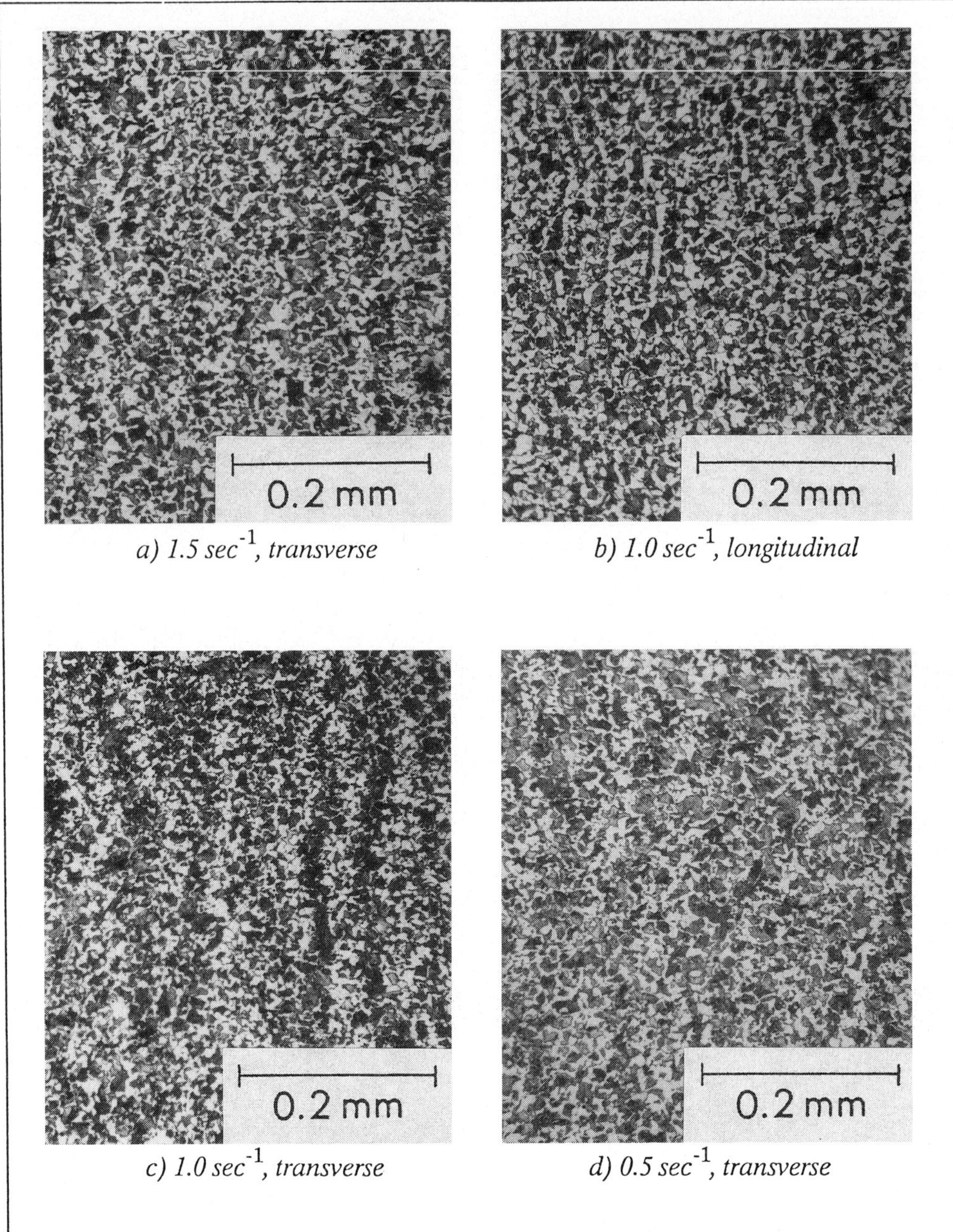

a) 1.5 sec^{-1}, transverse

b) 1.0 sec^{-1}, longitudinal

c) 1.0 sec^{-1}, transverse

d) 0.5 sec^{-1}, transverse

Figure 7 a-d: Compression specimens, 871°C

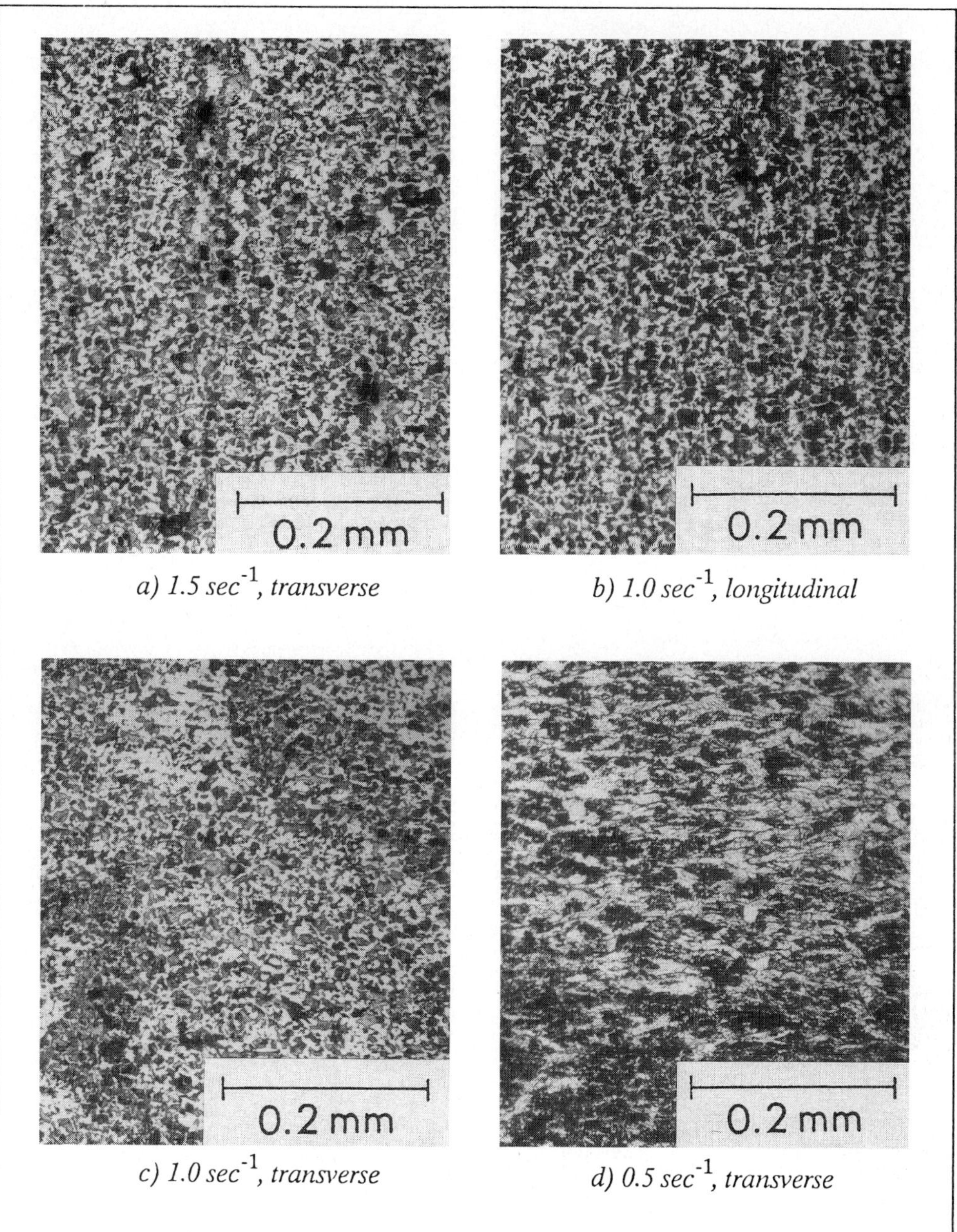

a) 1.5 sec^{-1}, transverse

b) 1.0 sec^{-1}, longitudinal

c) 1.0 sec^{-1}, transverse

d) 0.5 sec^{-1}, transverse

Figure 8 a-d: Compression specimens, 816°C.

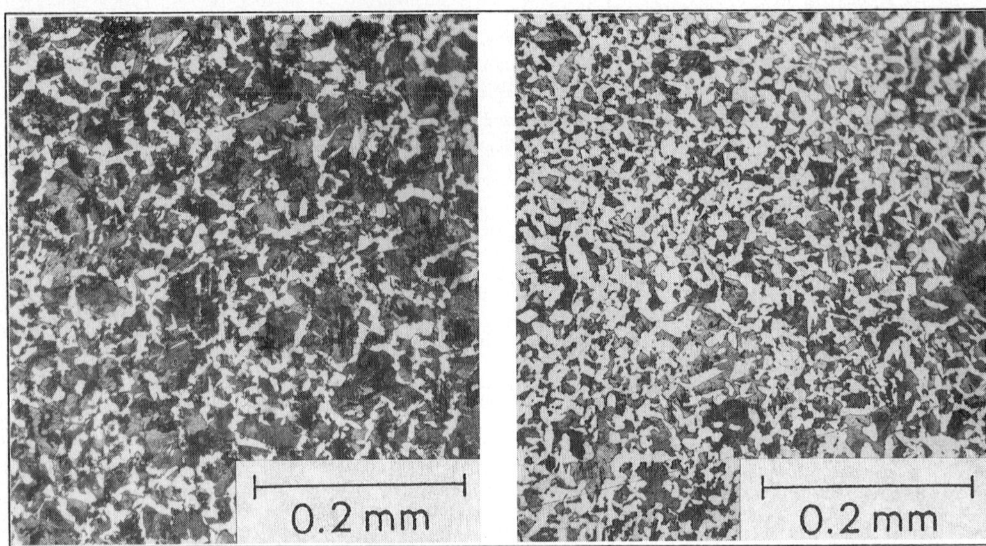

Figure 9: *Dead zone vs. statically heated, 982°C.*

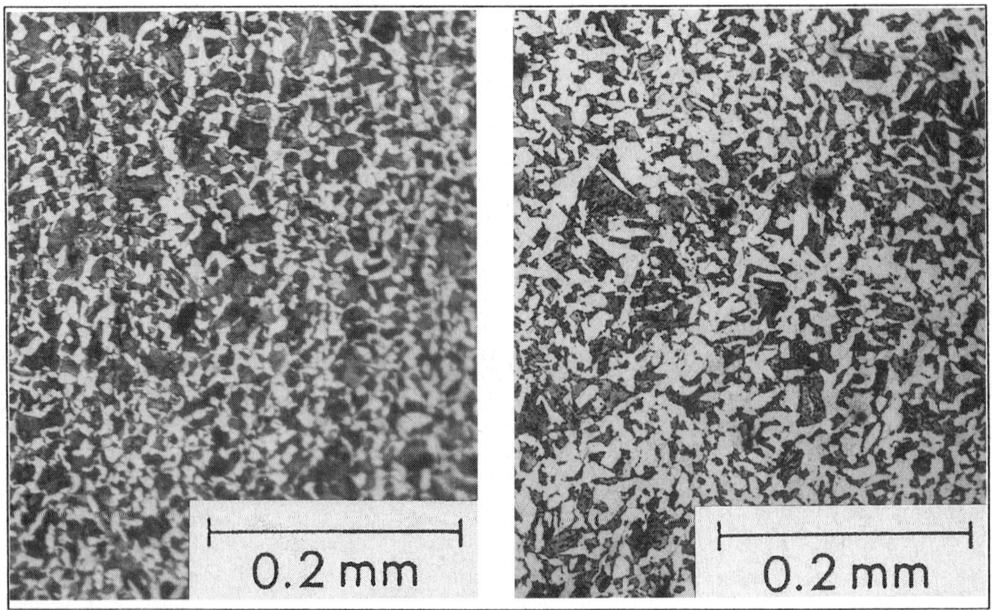

Figure 10: *Dead zone vs. statically heated, 927°C.*

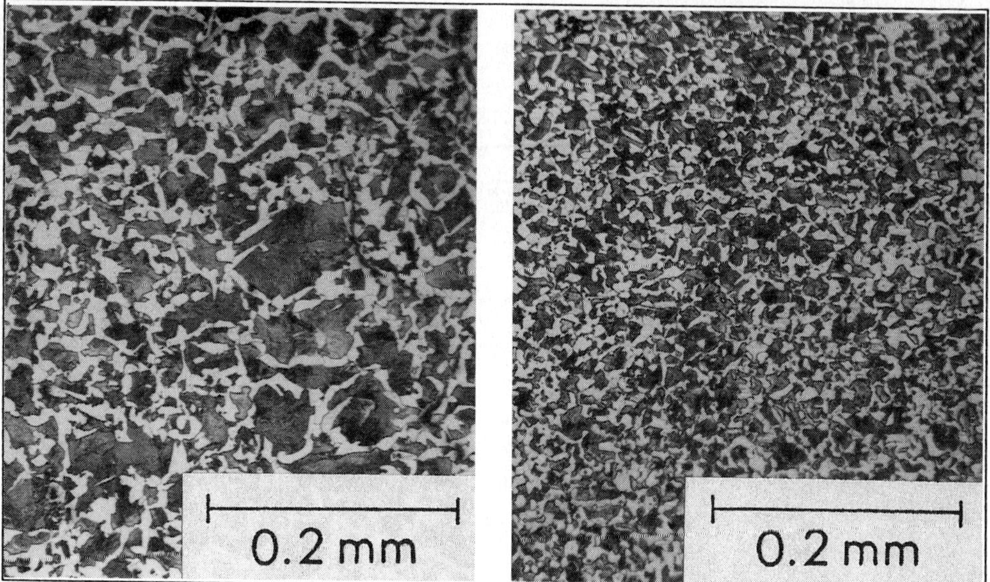

Figure 11: Constant strain rate vs. average strain rate, 871°C.

5. DISCUSSION AND RESULTS

Using light microscopy, the microstructure of hot-worked 1035 steel was found to vary with temperature and strain rate. Grain size increased with the increase in temperature of deformation. At high temperature, the strain rate affected the size distribution of pearlite and ferrite. Also, at the low temperature and strain rate, transformation occurred before the deformation ended.

In 1964, GRANGE [23] examined the microstructure alterations in steel after hot-rolling and demonstrated that lowering the hot-working temperature leads to reduced grain size, thus improving properties. The 1035 steel specimens in this study also demonstrated this temperature relationship with final grain size, although at higher temperatures a mixed grain structure appeared.

All the specimens were fully austenitic at the start of deformation but because of the different temperatures of deformation, there were differences in the time spent above the transformation temperature. At lower temperatures, the energy in the microstructure would be needed for the transformation, whereas at higher temperatures there would be energy available for recrystallization. This explains the mixed grain sizes seen at high temperature. During high temperature deformation, austenite grains differ in that some are newly recrystallized and others have varying amounts of work in them. After deformation, during slow cooling, the transformation occurs. Upon transformation the recrystallized grains become large areas of pearlite surrounded by smaller ferrite grains. The worked austenite grains, with many nucleation sites, transform into areas of medium sized pearlite and ferrite grains. In addition, there is carbon associated with the dislocations in the worked austenite, and the process of working the steel aids in the diffusion of carbon. To obtain this dynamically recrystallized microstructure, it is necessary to have the combination of high temperature and high strain rate.

The mechanism involved in the mixed size grain structure observed in the high temperature hot-worked 1035 specimens is most likely dynamic recrystallization. Although the specimens were hot-worked at a temperature in the low range of most investigations, the strain and strain rates fall within the range where dynamic recrystallization has been observed (Table IV).

The different areas of a compression specimen experience different dominating forces. In the central area, temperature effects are dominant with moderate amounts of work. In the shear areas near the dead zones mechanical work is more influential.

Excluding the dead zones, there was little variation in the different regions of the 1035 specimens until hot-working was done at the highest temperature. The large pearlite areas appeared in the regions where temperature and work were high. Large pearlite masses, which indicate recrystallization, were located in the central and shear regions of the high temperature specimens at the higher strain rates, again indicating the necessary combination of high temperature, high strain rate and a required minimum amount of strain. Also critical was the length of time spent above the transformation temperature. The recrystallization must have an incubation period for nucleation.

The specimens appeared similar longitudinal and transverse to the rolling direction. Upon hot-working some banding of the ferrite and pearlite appeared (Figure 6a-6d) suggesting some prior texture. In addition, barrelling of the specimens was not uniform, also providing evidence of inhomogeneity in the original plate which was not detectable with light microscopy. Differences in lubrication or asymmetry of loading may have contributed to the nonuniform barrelling. The type of barrelling and the banding were not associated with any particular testing condition.

The original investigators performed exploratory hot compression tests comparing constant strain rate tests with constant velocity (average strain rate) tests. The investigators reported no significant differences in the behavior of the steel under the two test conditions. Subsequent tests were done under constant velocity conditions. Although the flow stresses appeared similar, the microstructures showed some differences. The microstructures were similar only at the highest test temperature.

This study of hot-worked 1035 steel confirmed the observations of other investigators concerning the relationship between temperature and grain size. Working at lower temperatures will result in a finer grain size. This type of structure has more uniform and desirable properties.

At higher temperatures, a mixed grain structure can develop due to dynamic recrystallization. A strain rate effect was seen at the highest temperature tested, with recrystallization occurring at the higher strain rates. The mixed size grain structure, which appeared at regions in the specimens of high temperature and work, showed the necessary conditions for the onset of dynamic recrystallization in 1035 steel.

At the lowest temperature and strain rate, transformation occurred before the end of deformation. This is essentially a cold-worked structure which would require heat treatment for grain refinement.

For a fine grained product with uniform properties, 1035 steel should be hot-worked between 816°C and 927°C and between strain rates of 1.0 sec^{-1} and 1.5 sec^{-1}. At lower temperatures and strain rates, transformation will occur during deformation. At higher temperatures and strain rates dynamic recrystallization will result in a mixed size grain structure.

Dynamic Recovery		Dynamic Recrystallization
	Prerequisite Microstructure	
none,ideal:single phase with high SFE		none, ideal: pure single phase with low SFE
	Steady State Structure	
original grains elongated		recrystallized grains, mixed sizes larger grains elongated average grain size increases with increases in temperature or strain rate
equiaxed sub grains of constant size and misorientation dislocation density constant subboundaries decompose and reform subgrains larger and more perfect as temperature increases or strain rate decreases		dislocation substructure varies from none in nuclei to high density in larger grains average dislocation density constant
		same except repeated removal
	Texture	
develops with strain		same
	Final Microstructure	
elongated grains		fine grains, range of sizes few annealing twins
uniform dislocation substructure		dislocation density varies from grain to grain
SFE: stacking fault energy		

Table I: Summary of microstructures observed in dynamic recovery and recrystallization. (McQueen and Baudelet [6])

I. Supercritical TMT (above Ae_3)

a. Without recrystallization
(hot-cold work, high-temperature TMT)

b. With recrystallization
(controlled hot rolling, high temperature TMT)

II. Intercritical TMT (between Ae_3 and Ae_1)

III. Subcritical TMT (below Ae_1)

a. Prior to transformation
(ausforming, ausworking, ausrolling, low temperature TMT)

b. During transformation (isoforming)

c. After transformation

IV. Thermomechanical annealing

Table II: Classification of Thermomechanical Treatments (TMT) (Leslie [18])

GROUP	DYNAMIC	STATIC
A) A1, alpha iron, ferritic alloys 18/8 stainless at temperature less than 1000°C Other high SFE materials	Recovery (all strains)	Recovery followed by Recrystallization
B) Cu, Ni, gamma iron austenitic alloys Other low to moderate SFE materials	Recovery (small strains) Recrystallization (large strains)	Very limited recovery followed by recrystallization

Table III: Restoration Processes Associated with Hot-working (Tegart and Gittins [20])

	Strain	Strain Rate s^{-1}	Temperature °C	
Vodopirec et al.[24]	reduction 15-22% pass		900-1200 rolling	carbon steel recovery recrystallization above 1050°C
Semiatin and Holbrook [25]	.7	.009 .01 3.5 10.0 10.2	20-1200 compression torsion	austenitic stainless steel recrystallization at 1000°C and .01 s^{-1}
Sankar et al. [26]	.2 .4	.1 1.0 10.0	900-1000 rolling	low carbon steel recrystallization at .1 s^{-1}
McQueen et al. in [27]		2.3 23.0	800-1200 torsion	304 stainless steel recrystallization at 1100°C and 1200°C
Hardwick et al. in McQueen and Jonas [28]		.0011	1100	.25% carbon steel recrystallization

Table IV: Observations of Dynamic Recrystallization in Steels.

sec^{-1}	816°C	871°C	927°C	982°C
3.0	T	T	T	T,L
1.5	T,L	T,L	T,L	T,L
1.0	T	T	T	T
.5	C,A	C,A	C,A	C,A

T transverse to the rolling direction
L longitudinal to the rolling direction
C constant strain rate
A average strain rate

Table V: Summary of Test Conditions

6. REFERENCES

1. Jonas, J.J., Sellars, C.M., Tegart, W.J.McG, "Strength and Structure Under Hot-Working Conditions", Met Rev. Vol. 14, Editor: J. S. Bristow, The Institute of Metals, London, 1969.

2. Stüwe, H.P., "Do Metals Recrystallize During Hot Working?" Deformation Under Hot-Working Conditions, Editors: C.M. Sellars, W.J. McG. Tegart, The Iron and Steel Institute, Southend-on-Sea, 1968.

3. Jonas, J.J., McQueen, H.J., Wong, W.A., "Dynamic Recovery During the Extrusion of Aluminum", Deformation Under Hot Working Conditions, Editors: C.M. Sellars, W.J. McG. Tegart, The Iron Steel Institute, Southend-on-Sea, 1968.

4. McQueen, H.J., "Deformation Mechanisms in Hot Working", J. Metals, No. 20, April 1968, pp. 31-38.

5. McQueen, H.J., "Metallography and the Mechanisms Related to Hot Working of Metals", Microstructural Sci., Vol. 7, 1979, pp. 71-86.

6. McQueen, H.J., Baudelet, B., "Comparison and Contrast of Mechanisms, Microstructures, Ductilites in Superplasticity and Dynamic Recovery and Recrystallization", Strength of Metals and Alloys, Vol 1., Editors: P. Haasen, V. Gerold, G. Kostorz, Pergamon Press Ltd., Oxford, 1979.

7. Sakai, T. and Jonas, J.J., "A New Approach to Dynamic Recrystallization", in Deformation, Processing and Structure, Editor: G. Krauss, ASM, Metals Park, Ohio, 1984, pp. 185-228.

8. Pokorny, A., DeSerri Metallographia, Vol. III, Solidification and Deformation of Steels, The High Authority of the European Coal and Steel Community, Luxemburg, 1967.

9. Hoyt, S.L., ASME Handbook, Metals Properties, ASME, NY, 1954.

10. Carmichael, C., The Ferrous Metals Book, Editor: C. Carmichael, The Penton Publishing Co., NY, 1961.

11. Bardes, B.P., Metals Handbook, 9th ed., Vol. 1, Properties & Selection: Irons and Steels, ASM, Editor: B. P. Bardes, Metals Park, Ohio, 1078.

12. Teichert, E.J., Metallography and Heat-Treatment of Steel, Vol. 3, Ferrous Metallurgy, McGraw-Hill Book Company, Inc., NY, 1938.

13. Samuels, L.E., Optical Microscopy of Carbon Steels, ASM, Metals Park, Ohio, 1980.

14. McGannon, H.E, The Making, Shaping and Treating of Steel, Editor: H.E. McGannon, 8th ed., USS, 1964.

15. Baumeister, T., Standard Handbook for Mechanical Engineers, 7th ed., Eds., T. Baumeister and L.S. Marks, McGraw-Hill Book Company, 1958.

16. Shewmon, P.G., Transformations in Metals, McGraw-Hill, Inc., NY 1969.

17. Michalak, J.T., "Plastic Deformation Structures in Iron and Steel", Metals Handbook, 8th ed., Vol. 8, Metallography, Structures and Phase Diagrams, ASM, 1961.

18. Leslie, W.C., The Physical Metallurgy of Steels, Hemisphere Publishing Corporation, Washington, 1981.

19. Jolley, W. and Witmer, D.A., "Recovery, Recrystallization and Grain Growth Structures in Iron and Steel", Metals Handbook, Vol. 8, ASM, Metals Park, Ohio, 1961.

20. Tegart, W.J. McG. and Gittins, A., "The Hot Deformation of Austenite", in The Hot Deformation of Austenite , J.D. Ballance, AIME, NY, 1977.

21. Brown, E.L., DeArdo, A.J., The Behavior of Austenite During Hot Rolling", Strength of Metals and Alloys, Vol. 1, Eds., P. Haasen, V. Gerold, G. Kostorz, Pergamon Press, Ltd., Oxford, 1979.

22. Bernshtein, M.L., Kaputkina, L.M., Prokoshkin, S.D., Dobatkin, S.V., "Structural Changes During Hot Deformation of Austenite in Alloy Steels", Acta Metall., Vol. 33, No. 2, Feb. 1985, pp. 247-254.

23. Grange, R.A, "Microstructural Alterations in Iron and Steel During Hot Working", Fundamentals of Deformation Processing, Editor: W.A. Backofen, Syracuse University Press, Syracuse, 1964.

24. Vodopirec, F., Gabrovsek, M., Kmetic, M., Rodic, A., "Interpass Recrystallization of Austenite in Some Steels During Rolling", Met Tech., Vol. 11, Nov. 1984, pp. 481-488.

25. Semiatin, S.L. and Holbrook, J.H., "Plastic Flow Phenomenology of 304L Stainless Steel", Met. Trans. A., Vol. 14A, 1983, pp. 1681-1695.

26. Sankar, J., Hawkins, D. and McQueen, H.J., "Behavior of Low Carbon and HSLA Steels During Torsion-Simulated Continuous and Interrupted Hot Rolling Practice", Met. Techn., Vol. 9, 1979, pp. 325-331.

27. McQueen, H.J., Petkovic, R., Weiss, H. and Hinlon, L.G., "Flow Stress and Microstructural Changes in Austenitic Stainless Steel During Hot Deformation", in The Hot Deformation of Austenite, Editor: J.B. Ballance, AIME, NY., 1977.

28. McQueen, H.J. and Jonas, J.J, "Recovery and Recrystallization During High Temperature Deformation", in Treatise on Materials Science and Technology, Vol. 6, Editor: R.J. Arsenault, Academic Press, NY., 1975.

Chapter 4

Computer-Aided Analysis and Modelling of Plastic Behaviour of Steels at Elevated Temperatures

by B.Kaftanoglu

1. ABSTRACT

The objective of this study is to develop a computer aided model of plastic properties of the materials as a function of stress, strain, strain-rate and temperature. Such a tool is expected to provide adequate input to the engineer or materials scientist to describe and examine the material behaviour in the modern manufacturing life. In order to realize this goal, experiments both at elevated temperatures and ambient conditions were conducted with 0.02% Nb and 1035 steels for various strain rates on a compression press. The model coefficients of some proposed constitutive equations were determined through an interactive computer program employing both optimum data fitting and computer graphics techniques. In order to further assess and exhibit the material behaviour, Bezier and B-spline curve and surface generation techniques were employed on a graphics capable computer. Using these modelling techniques, a designer can interactively develop a material model by using experimental data in both 2D and 3D models.

2D and 3D representations using experimental data proved to be successful with respect to the choice of constitutive equations and methodology. The paper covers the basic theory and experiments and describes the modelling studies. The developed material model's application to hot and cold rolling of flat strip, roll forming and tube-drawing is also referred to.

LIST OF SYMBOLS

A	area
A_1, B_1, A_2	constants in empirical constitutive equations
B	tensor quantity including position vectors of each vertex
c	constant equal to strain-rate
d	sample diameter
F(x)	Lagrange interpolation function
h	distance in thickness direction
$JB_{n,i}(t^*)$	Bernstein Function
K	order of B-Spline curve
$K_{m,j}(w)$	Bernstein Function
L	load or external force
I	length
ln	natural logarithm
log	logarithm to base 10
m	work softening exponent
mm	order of polynomial
$M_{j,l}$	weighting function of B-Spline surfaces
n	work hardening exponent
nn	order of polynomial
$N_{i,k}(t^*)$	weighting function of B-Spline curves
P_i	vector components of vertices
$P(t^*)$	parametric function of curves
P(u,w)	parametric function of surfaces
Q(u,w)	parametric function of surfaces
r	radial distance
t^*	parameter
T	absolute temperature

T_x	transformation matrix about x-axis
T_y	transformation matrix about y-axis
u	parameter
V	instantaneous ram speed
w	parameter
X	knot vector
Y	knot vector

Greek Symbols

ε	natural or true strain
$\bar{\varepsilon}$	equivalent strain
$\dot{\varepsilon}$	true strain-rate
σ	normal stress; true stress
$\bar{\sigma}$	equivalent stress
σ_m	true ultimate tensile strength
$\sigma_1, \sigma_2, \sigma_3$	principal stresses
$d\bar{\varepsilon}$	plastic strain increment
$\hat{\sigma}_i$	calculated flow stress
σ_i	experimental flow stress

Indices

i	vertex index
j	vertex index
0	original

2. INTRODUCTION

Plastic properties of materials are needed for theoretical modelling of metal forming processes. Using plasticity theory in the development of theoretical models, relationship between the equivalent-stress and equivalent-strain must be known. This relationship can be determined using carefully conducted experiments under controlled conditions. Since this relationship is sensitive to temperature and strain-rate, tests need to be carried out in a range of these parameters representative of the metal forming process being investigated. These material properties at room and elevated temperatures are needed for processes such as rolling, extrusion, forging and other bulk and sheet forming applications.

JOHNSON and MELLOR [1] discussed a number of constitutive equations used to model plastic properties of materials at room and elevated temperatures. KAFTANOGLU and SIVACI [2] investigated the properties of certain ferrous and nonferrous materials at elevated temperatures at low strain-rates and used nonlinear regression analysis to fit experimental data with a Swift type constitutive equation. Rate equations have also been used to express the kinetics of hot deformation by JONAS et al. [3,4].

Modern manufacturing technology has increasingly many applications in complex boundary value problems of plastic deformation, which must be solved with high precision requiring the use of constitutive relations. Knowledge of flow and stress in metal forming operations is of great importance in determining the optimum forming conditions and predicting the final properties of the workpiece.

The connection among externally measurable parameters such as stress, strain, strain-rate, temperature and some internal parameters such as hardness, which govern the plastic behaviour of the solids, is called a constitutive equation.

Constants of the constitutive equations are always derived from experimental results and the validity of the equations are justified by comparing them with the experimental values. The most common method for developing constitutive equations is the optimization of the constants in the proposed equations by using experimental data. Although they are generally useful in the range of parameters where actual measurements were made, they are often inadequate for extrapolation.

The description of the stress-strain curves and strain-hardening of metals by a mathematical expression is a frequently used approach. This is the easiest way of formulating metal forming problems, and designers always enjoy to work with the analytical equations. Although these formulations are generally suitable for many purposes, by introducing new constants (by optimizing experimental values or other techniques) one cannot generalize the validity of these equations for every material.

AISI 1035 steel was tested at 815°C-982°C (1500°F-1800°F) range at a strain rate varied between 0.5-1.5 sec^{-1} [5]. HSLA steel (0.225 in = 5.7 mm) with 0.02% Nb composition was tested both at the room temperature and in the temperature range of 750-950°C for a nearly constant strain-rate of 13.3 sec^{-1} [6]. KADIOGLU [7] reviewed previous studies at high temperature and strain-rate conditions and investigated effects of different parameters. He developed an optimum data fitting model with respect to constitutive equations using computer graphics. KARAYAKA [8] developed a computer aided model of the plastic properties of the materials as a function of stress, strain, strain-rate, and temperature. He employed Bezier and B-spline curve and surface generation techniques. He used KAFTANOGLU and HOW-RANG ONG's data [5]. YÜRÜR [6] used his own data in order to provide true stress true strain curves for niobium steel at different temperatures. He used KADIOGLU's [7] program for model fitting to this set of data. He found reasonable agreement especially at lower temperatures.

3. THEORY

Tensile, compressive or torsion tests may be used to determine the metal's properties in the plastic range. It is of importance to choose the type of test so that it broadly simulates the forming process which is to be modelled.

In the present investigation a compressive test is chosen since the applications intended are rolling, forging and tube drawing which are predominantly compressive in nature.

A typical specimen for the compression test is shown in Fig. 1.

For this specimen:

$$\text{Compressive true strain} = \varepsilon = \ln \frac{h}{h_0} \qquad (1)$$

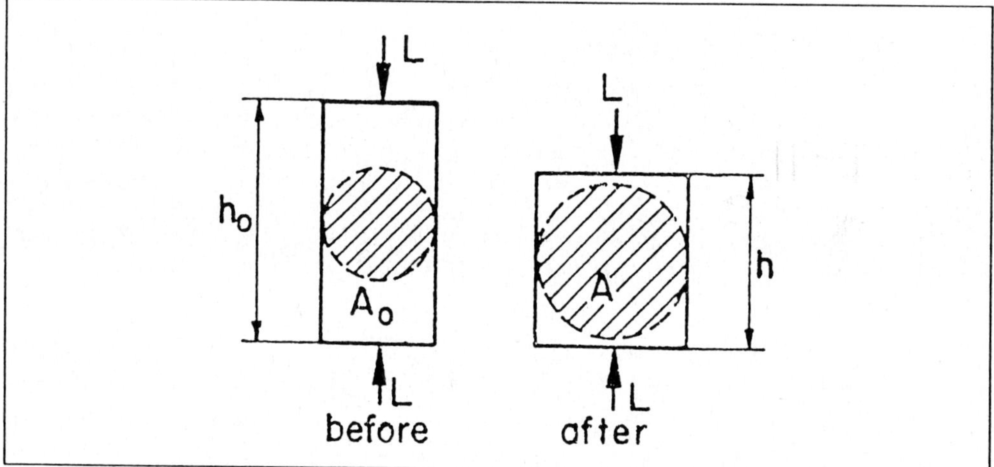

Figure 1: Specimen for Compressive Test.

Due to the constancy of volume

$$A_0h_0 = Ah$$

and

$$A = \frac{A_0h_0}{h} \tag{2}$$

The stress in the plastic range is defined by:

$$\sigma = \frac{Load}{A} = \frac{L}{A} \tag{3}$$

For experiments conducted at constant strain-rate, strain-rate is defined as:

$$\dot{\varepsilon} = \frac{d\varepsilon}{dt} = \frac{d}{dt}\left(\ln\frac{h}{h_0}\right) = constant = c \tag{4}$$

where $c = \frac{1}{h}\frac{dh}{dt} = \frac{V}{h}$ (5)

also $h = h_0e^{ct}$ (6)

Based on the MISES [1] theory, the equivalent stress is defined as:

$$\bar{\sigma} = \frac{1}{\sqrt{2}}\left[(\sigma_1 - \sigma_2)^2 + (\sigma_2 - \sigma_3)^2 + (\sigma_3 - \sigma_1)^2\right]^{\frac{1}{2}} \tag{7}$$

and the equivalent plastic strain-increment is defined as:

$$d\bar{\varepsilon} = \frac{\sqrt{2}}{3}\left[(d\varepsilon_1)^2 + (d\varepsilon_2)^2 + (d\varepsilon_3)^2\right]^{\frac{1}{2}} \tag{8}$$

The relationship between $\bar{\varepsilon} = \int_0^{\varepsilon} d\bar{\varepsilon}$ has to be obtained from experiments and needs to be expressed using an empirical formula. A new empirical equation which is proposed is given below which allows for work hardening and work softening:

$$\bar{\sigma} = A_1(B_1 + \bar{\varepsilon})^n - A_2(\bar{\varepsilon})^m \tag{9}$$

The first part of this equation expresses the work hardening and the second part expresses the work softening. A_1 and A_2 are scaling constants for the respective parts.

B_1 expresses the initial strain for cold-work, n is the work-hardening exponent and m is the work-softening exponent.

Another well-known empirical relation is SWIFT's equation:

$$\bar{\sigma} = K(\varepsilon_0 + \bar{\varepsilon})^n \tag{10}$$

where ε_0 expresses the prior strain history and K and n are the material constants.

LUDWIK proposed an empirical equation in the form:

$$\bar{\sigma} = A + B(\bar{\varepsilon})^m \tag{11}$$

and a power law of the form:

$$\bar{\sigma} = K(\bar{\varepsilon})^n \tag{12}$$

where A, B, K, m, and n are the material constants [9].

VOCE proposed an exponential equation in the form:

$$\bar{\sigma} = A - (A - B)\exp(-C\bar{\varepsilon}) \tag{13}$$

where A, B, C are the material constants [10].

4. EXPERIMENTS

Experiments have been conducted on cylindrical specimens with 3.8 cm in height and 1.27 cm in diameter. They were first machined and wrapped in ceramic wool for insulation against heat loss. A 25 ton closed loop controlled MTS testing machine has been used with specially made tungsten carbide platens. A special electronic circuit has been designed to achieve constant strain-rate tests.

The specimens insulated with ceramic wool have been heated in a furnace at a temperature higher than the test temperature and tests were started at a precalibrated time such that the temperature of the specimen dropped down to the desired test temperature. The tests were carried out using the MTS machine at the required strain-rates. Load versus displacement data was recorded using a digital recorder. This set-up was used for testing AISI 1035 steel at elevated temperatures [5]. A second set-up was developed and used by YÜRÜR [6]. He tested primarily niobium steel.

The experiments were carried out in a mechanical eccentric crank press. The specifications are:

Load	: 60 tons
Strokes per min.	: 55 strokes/min
Stroke adjustment	: 0-100 mm
Connecting rod movement	: 0-80 mm
Motor power	: 5.5 Hp
Motor speed	:1400 rpm

The experimental set up is shown in Figure 2.

Load cells have been designed and manufactured to determine the instantaneous applied pressure during the tests. Two types, with the maximum measuring capacity of 2 and 3 tons have been manufactured. In the cells HBM 6/120LY61 type strain gages have been employed. Displacements were measured by appropriate WDT's with a precision of about ± 0.003 mm/mm.

In order to decrease the effect of friction and prevent the instability against the lateral shear, cylindrical slender specimens have been cut both at the lateral and longitudinal directions, and the slenderness ratio has been kept at 1.5. Consequently, for the second set of experiments specimens with 7.5 mm in height and 5 mm in diameter have been machined and the surfaces have been ground.

For insulation against the heat loss at the elevated temperatures, the lateral surfaces of the samples have been covered with 3.5 mm thick clay mud. Proper concentric covering has been accomplished by a die. However, the two ends of the specimens were let uncovered to provide suitable contact of the die and the test piece during the tests.

In order to avoid or decrease barrelling of the specimens, the frictional effects of the matching surfaces were tried to be minimized by means of lubrication. For this purpose, Molykote 321 R lubricant was used.

Set-up is equipped with a HP 9816 processor using a BASIC 3.0 software. A Lindberg 51894 model 3000 watt $\pm 5°C$ furnace was used to heat the specimens. A temperature measurement set-up with 0.5°C accuracy in the 0-1200°C range was incorporated.

5. MODELLING

The experimental data recorded has been processed by a computer program. Simply stating, a curve is fitted to the stress-strain or stress-strain-strain rate data employing an optimization algorithm. KAFTANOGLU and ONG [5] used the following objective function:

$$\text{Error} = S = \sum_{i=1}^{p} [\sigma_i - A_1(B_1 + \bar{\varepsilon}_i)^n + A_2(\bar{\varepsilon}_i)^m]^2 \tag{14}$$

where $S = f(A_1, B_1, n, A_2, m)$

and

σ_i and ε_i are experimental points and
p is their number.

Using this method, the least squares error between the constitutive equation and the experimental data is minimized and the values of A, B_1, n, A_2 and m are calculated.

Using the values of the material constants for each strain-rate and temperature, a computer plot is obtained for σ - ε relationship showing experimental points and the fitted curve.

For each temperature level three-dimensional computer plots have also been obtained where the true-stress is shown as function of true-strain and strain-rate.

KADIOGLU [7] extended this approach to other models using MARQUARDT's [11] optimum interpolation between Taylor series method and steepest descent method. First the model is linearized by expanding σ_i in Taylor series about current trial values for the coefficients of the model. He treated the following models:

$$\bar{\sigma} = A_1(B + \bar{\varepsilon})^n - A_2(\bar{\varepsilon})^m \tag{15.1}$$

$$\bar{\sigma} = A_1(B + \bar{\varepsilon})^n \tag{15.2}$$

$$\bar{\sigma} = A_1 + B(\bar{\varepsilon})^n \tag{15.3}$$

$$\bar{\sigma} = A_1 - (A_1 - A_2)\exp(-A_3\bar{\varepsilon}) \tag{15.4}$$

In this study, to describe the stress/strain behaviour of materials a new constitutive equation is also proposed. The equation has been thought to represent adequately three different phenomena. These are work hardening, work softening, and history prior to strain.

The first two phenomena are described with the well known power law using a different constant for each. The third part is assumed constant for a specific temperature and strain rate. The new equation is given in the form:

$$\sigma = A_1(\varepsilon)^{A_2} - A_3(\varepsilon)^{A_4} + A_5 \tag{15.5}$$

where A1, A2, A3, A4, and A5 are material constants.

The objective function which is the summation of the square of residuals is in general:

$$S = \sum_{i=1}^{N} \left(\sigma_i - \hat{\sigma}_i\right)^2 \tag{15.6}$$

where σ is the experimental stress data and σ is the fitted stress point. N is the number of data points.

KARAYAKA [8] applied Bezier curves, surfaces and B-spline curves and surfaces to KAFTANOGLU's AISI 1035 data [5].

5.1 Bezier Curves [12]

The mathematical basis of the Bezier technique is a polynomial blending function which interpolates between the first and the last vertices. The Bezier polynomial is related to the Bernstein polynomial. Thus, the Bezier curve is said to have a Bernstein basis. The basis is given by:

$$JB_{nn,i}(t^*) = \frac{i!}{i!(nn - i)!} t^{*i}(1 - t^*)^{nn-i} \tag{16}$$

where nn is the degree of the polynomial and i is the particular vertex in the ordered set (0 to n). The curve points are given by:

$$P(t^*) = \sum_{i=0}^{nn} P_i JB_{nn,i}(t^*) \qquad 0 \le t^* \le 1 \tag{17}$$

where Pi contains the vector components of the vertices. Coordinates of a point on the curve can be calculated independently by:

$$\text{x-coordinate } P_x(t^*) = \sum_{i=0}^{nn} P_{ix} JB_{nn,i}(t^*)$$
$$\text{y-coordinate } P_y(t^*) = \sum_{i=0}^{nn} P_{iy} JB_{nn,i}(t^*) \qquad (18)$$
$$\text{z-coordinate } P_z(t^*) = \sum_{i=0}^{nn} P_{iz} JB_{nn,i}(t^*)$$

This property is used to develop an optimization algorithm.

5.2 Bezier Surfaces

The Bezier surfaces are extensions of the Bezier curves. The Bezier surfaces can be represented in the form of a cartesian product surface in terms of parametric variables u and w. The surface is given by:

$$Q(u,w) = \sum_{i=0}^{nn} \sum_{j=0}^{mm} B_{i+1,j+1} JB_{nn,i}(u) K_{mm,j}(w) \qquad (19)$$

where by analogy with equation (16),

$$JB_{nn,i}(u) = \frac{nn!}{i!(nn-i)!} u^i (1-u)^{nn-i} \qquad (20)$$
$$K_{mm,j}(w) = \frac{mm!}{j!(mm-j)!} w^j (1-w)^{mm-j}$$

and nn and mm are one less than the number of polygon vertices in the u and w directions respectively.

The B-tensor is composed of position vectors of the data.

5.3 B-spline Curves

B-spline basis contains the Bernstein basis as a special case. This basis is generally nonglobal. The nonglobal behaviour of the B-spline curves is due to the fact that each vertex Pi is associated with a unique basis function. Thus, each vertex P affects the shape of a curve over a range of parameter values. The B-spline basis also allows the order of the resulting curve to be changed without changing the number of defining polygon vertices.

A curve generated by B-spline basis is given by:

$$P(t) = \sum_{i=0}^{nn} P_i N_{i,k}(t^*) \tag{21}$$

where P_i are the defining polygon vertices.

Weighting function $N_{i,k}(t^*)$ are defined by the recursion formulas:

$$N_{i,1}(t^*) = \begin{cases} 1 \;\; if \;\; X_i \leq t^* < X_{i-1} \\ 0 \;\; otherwise \end{cases}$$

$$N_{i,k}(t^*) = \frac{(t^* - X_i)N_{i,k-1}(t^*)}{X_{i+k-1} - X_i} + \frac{(X_{i-k} - t^*)N_{i+1,k-1}(t^*)}{X_{i+k} - X_{i+1}} \tag{22}$$

The values of X_i are elements of a knot vector. A knot vector is simply a series of the real integers X_i, such that $X_i \leq X_{i-1}$ for all X_i. Examples of knot vectors are [0 1 2 3 4] and [000 0 1 1 2 3 3 3]. This vector also specifies the variation of parameter t which is not restricted to be between 0 and 1 as in the Bezier technique.

5.4 B-spline Surfaces

Implementation of B-spline surfaces can take many forms. The simplest form is the cartesian product surface analogous to the Bezier surface. For a B-spline surface the corresponding cartesian product surface is:

$$Q(u,w) = \sum_{i=0}^{nn} \sum_{j=0}^{mm} B_{i+1,j+1} N_{i,k}(u) M_{j,1}(w) \tag{23}$$

where by analogy with equation (19) $N_{i,k}(u)$ and $M_{j,1}(w)$ can be defined in similar manner. Thc $B_{i+1,j+1}$ are the position vectors of the defining polygonal surface.

As with B-spline curves, knot vectors can be defined in either u or w directions.

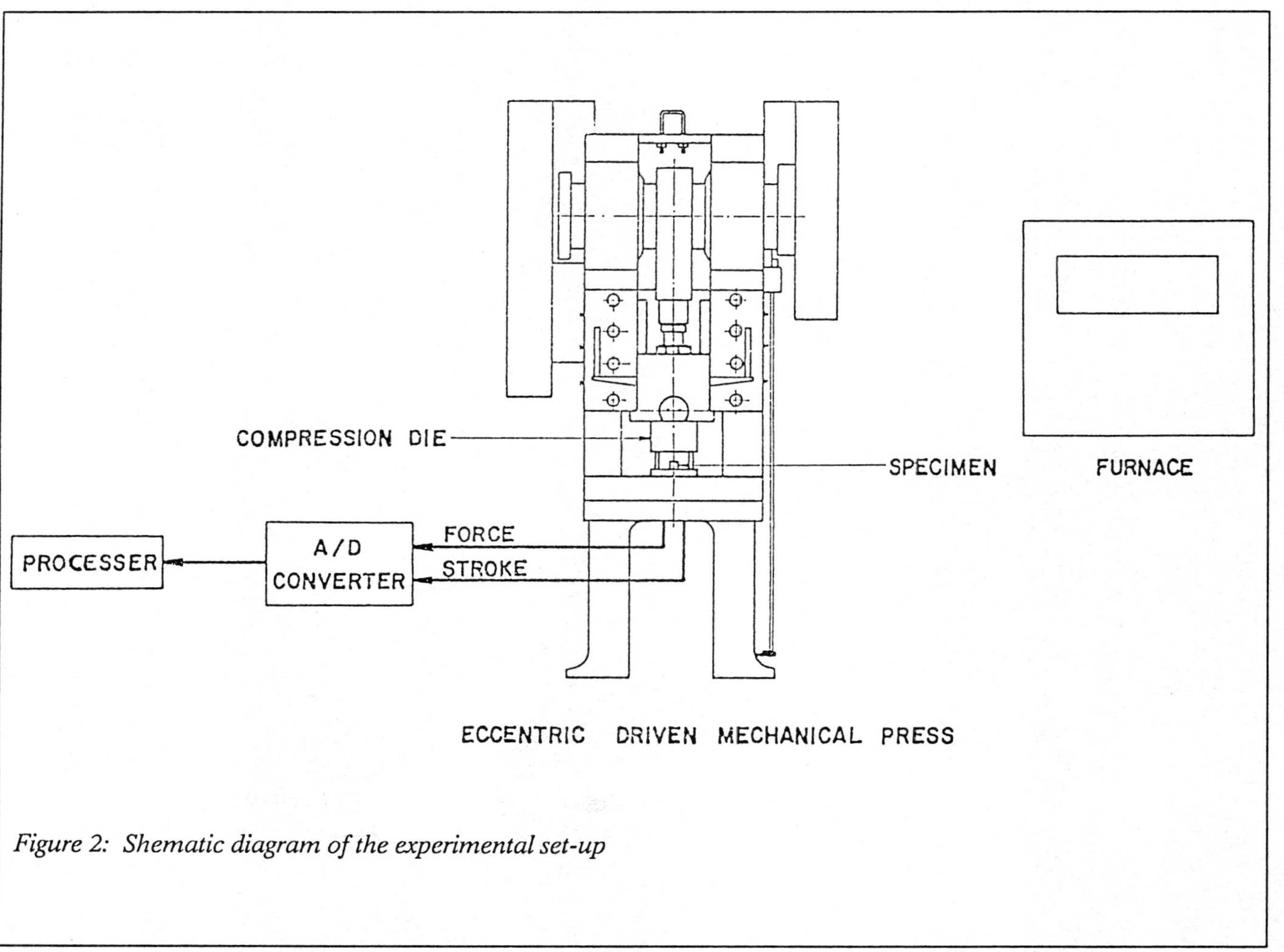

Figure 2: Shematic diagram of the experimental set-up

TABLE-1: MATERIAL CONSTANTS IN THE PLASTIC RANGE

Test Temp. °F/°C	Strain Rate sec^{-1}	A_1		A_2		B_1	n	m
		psi	MPa	psi	MPa			
	0.5	92718	639	76249	525	0.044	0.509	1.388
1500°F	1.0	100423	692	91404	630	0.048	0.567	1.422
816°C	1.5	126384	871	184673	1272	0.139	0.969	1.889
	0.5	116128	800	94952	654	0.015	0.542	0.911
1600°F	1.0	103796	715	83943	578	0.0121	0.510	0.903
871°C	1.5	137858	950	203682	1403	0.093	0.970	1.813
	0.5	100885	695	91528	631	0.007	0.522	0.807
1700°F	1.0	92341	636	80982	558	0.022	0,586	1.024
927°C	1.5	66092	455	116141	800	0.152	0.966	2.226
	0.5	98702	680	99612	686	0.009	0.623	0.925
1800°F	1.0	104166	718	94017	648	0.015	0.698	0.975
982°C	1.5	69608	480	88035	607	0.106	0.966	1.773

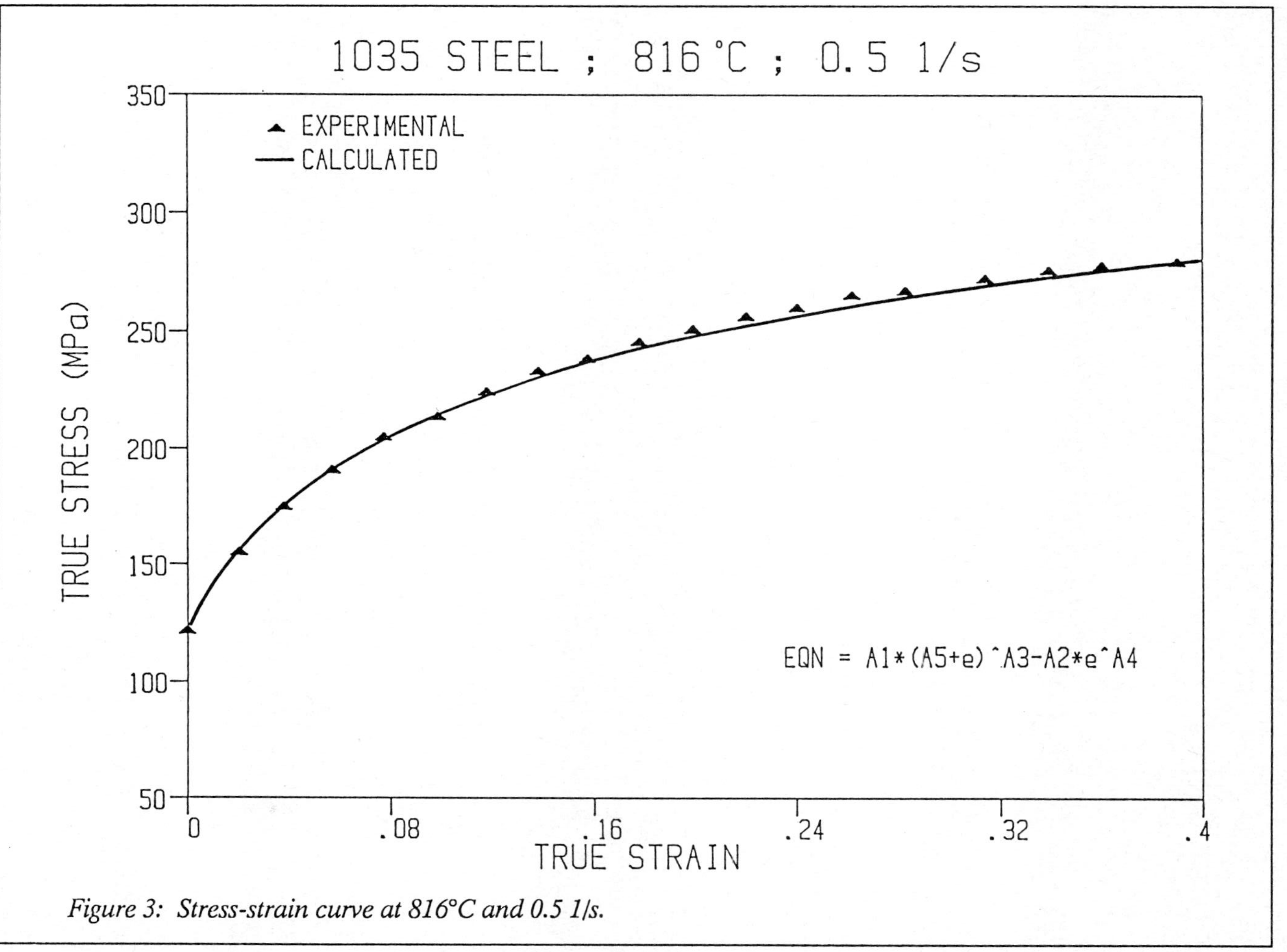

Figure 3: Stress-strain curve at 816°C and 0.5 1/s.

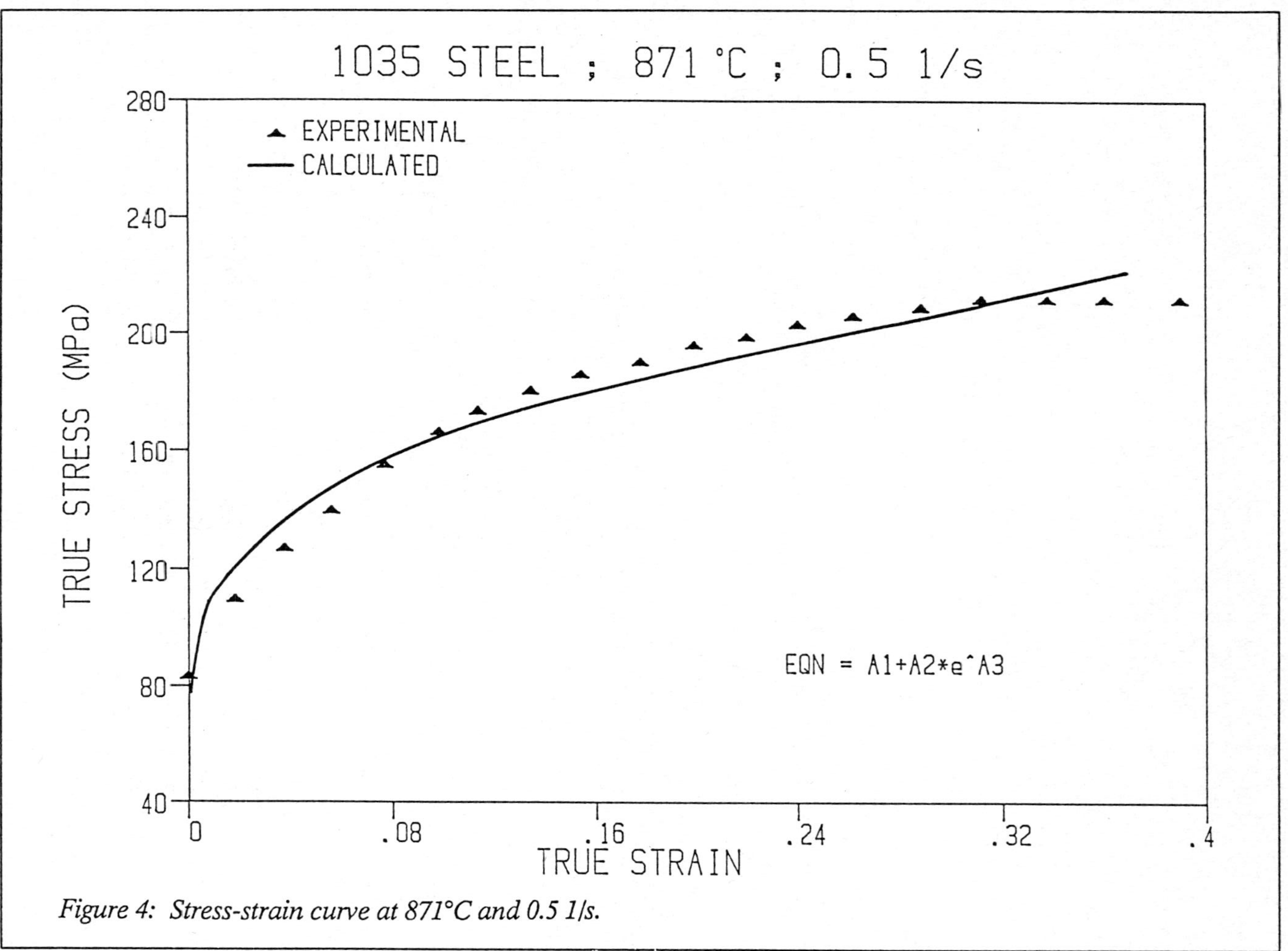

Figure 4: Stress-strain curve at 871°C and 0.5 1/s.

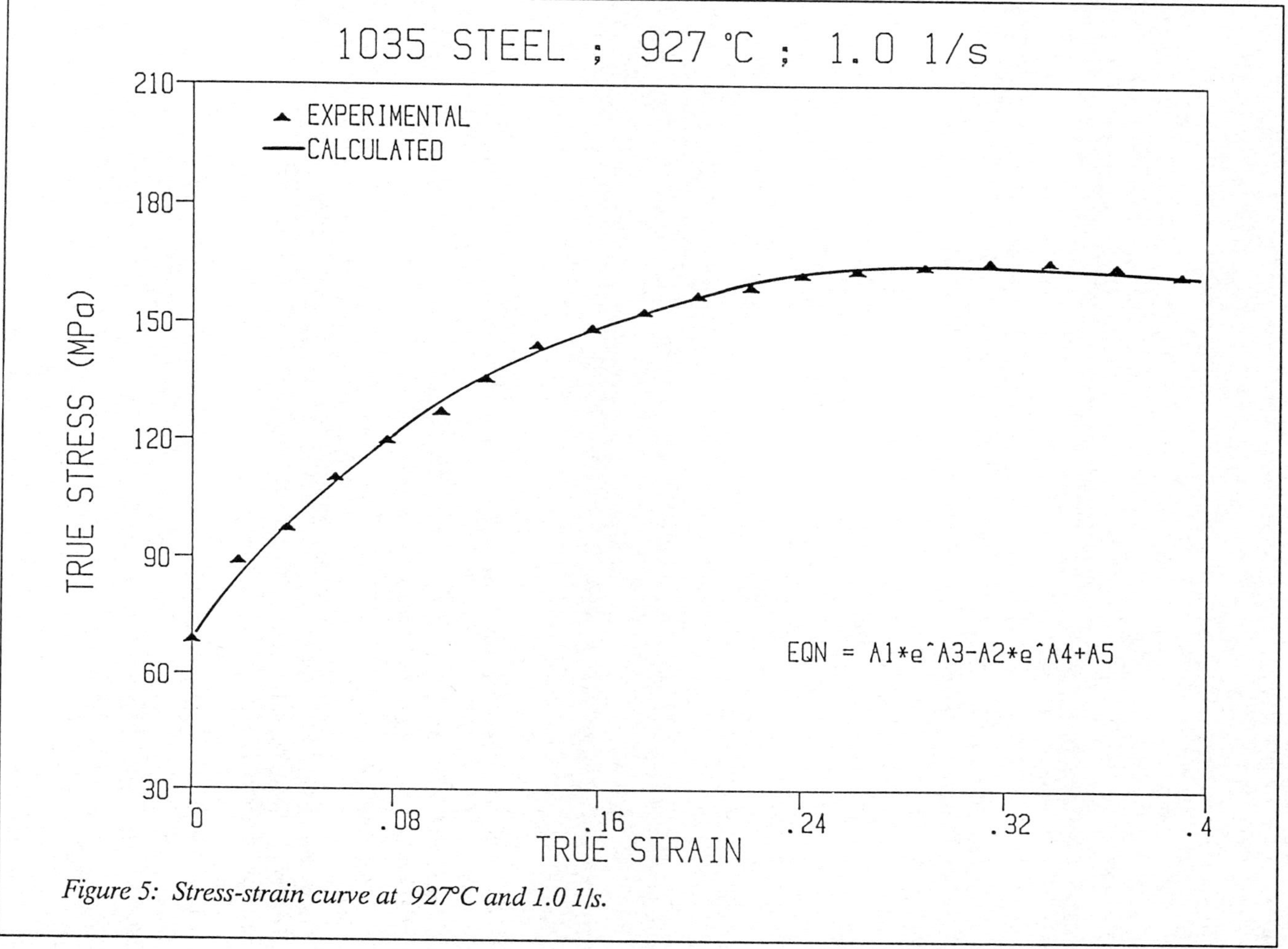

Figure 5: Stress-strain curve at 927°C and 1.0 1/s.

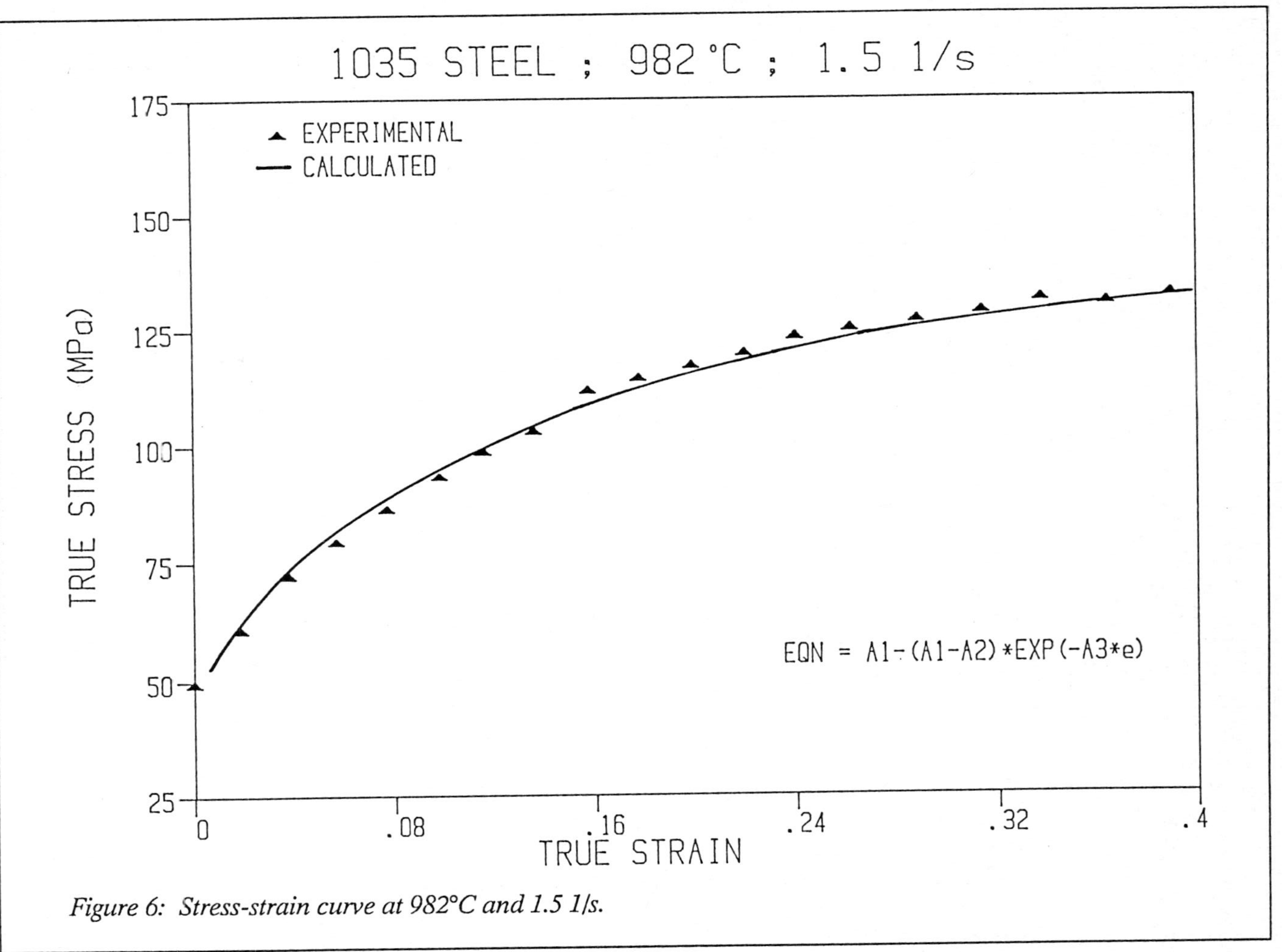

Figure 6: Stress-strain curve at 982°C and 1.5 1/s.

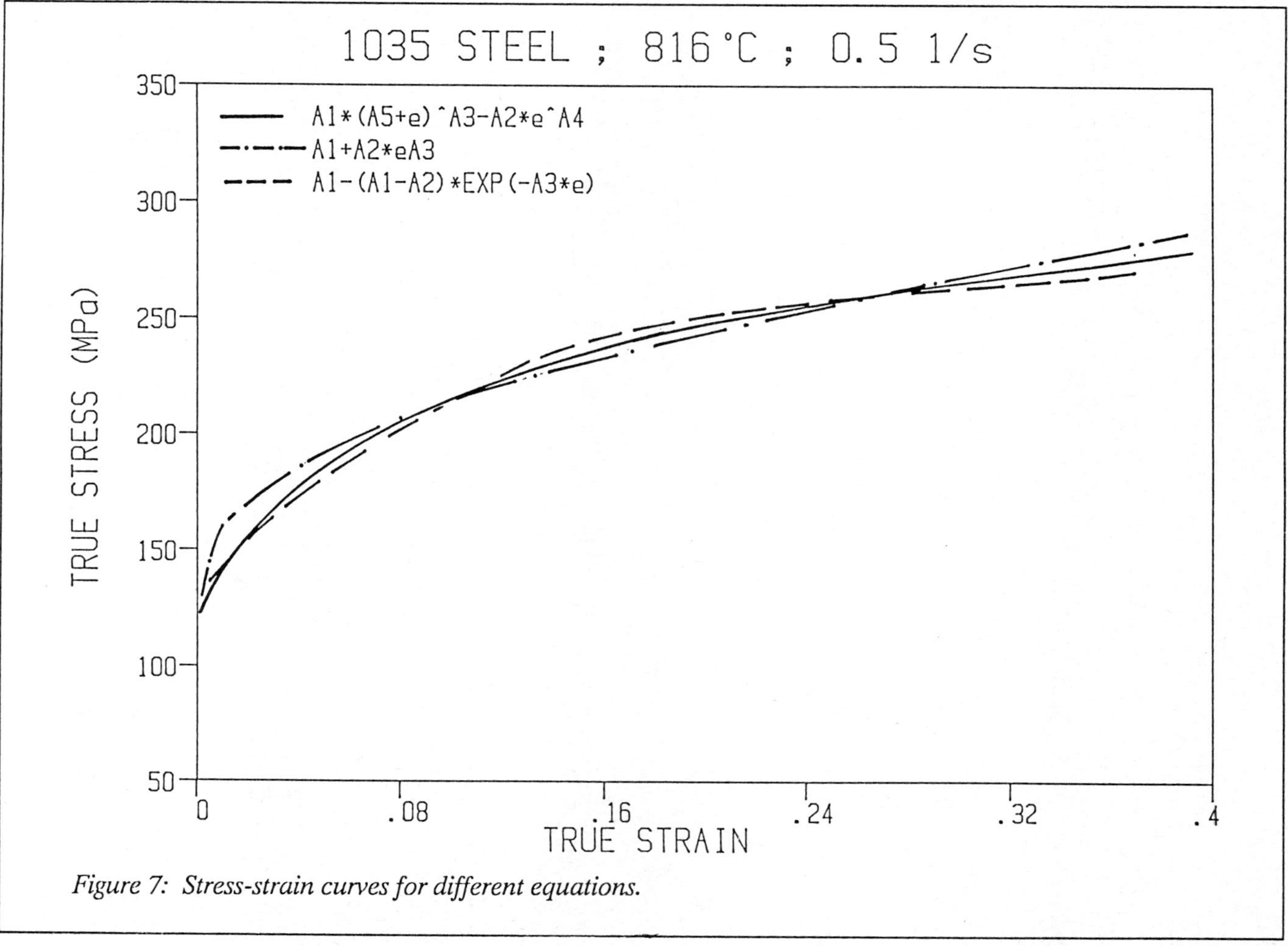

Figure 7: Stress-strain curves for different equations.

1035 STEEL ; 816 °C ; 0.5 1/s
EXPERIMENTAL
FIRST BEZIER FIT
OPTIMIZED BEZIER FIT
TRUE STRESS (MPa)
315
270
225
180
135
90
45
TRUE STRAIN
0
.08
.16
.24
.32
.4

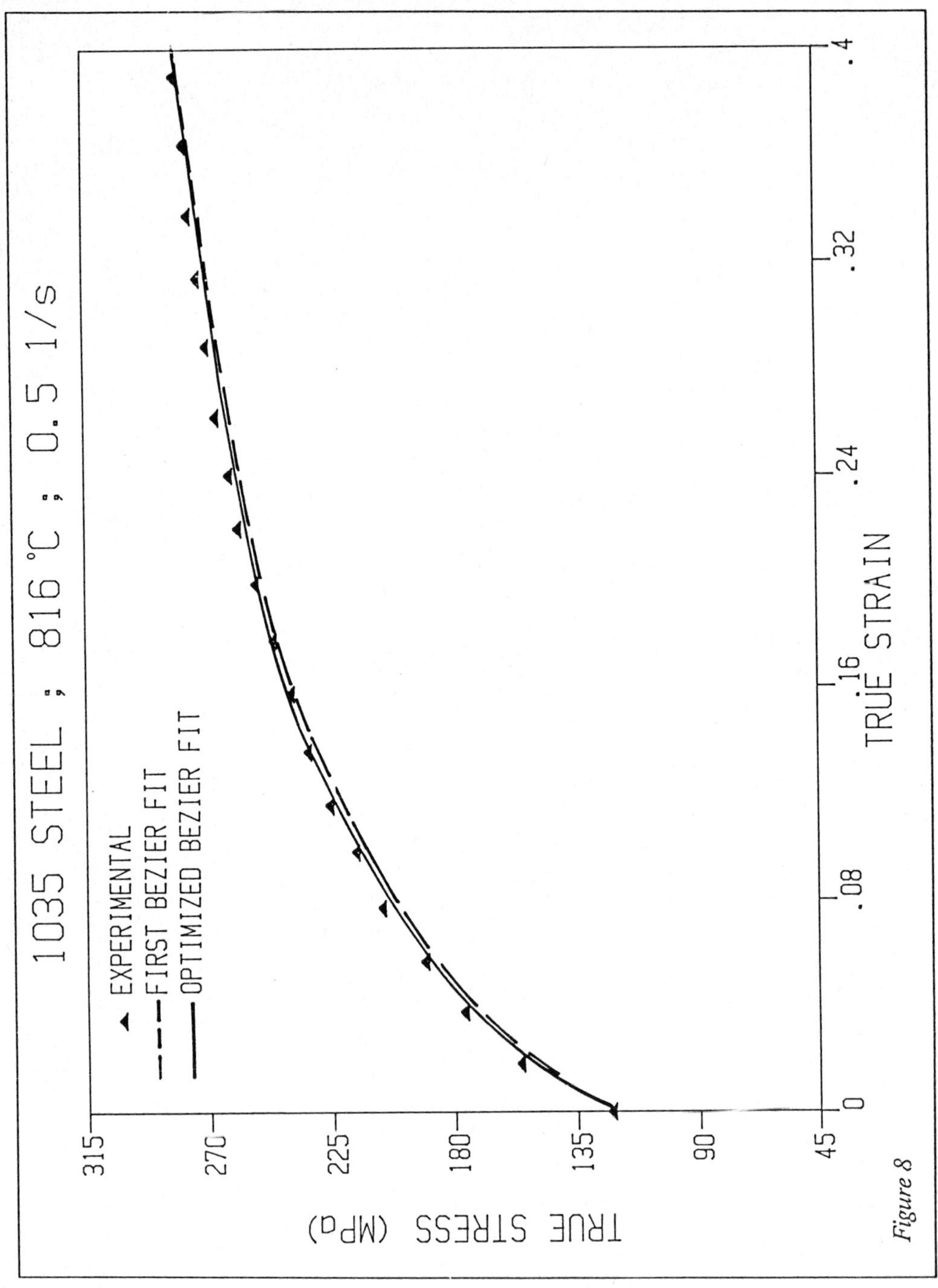

Figure 8

1035 STEEL
816 °C
FLOW STRESS (MPa)
280
226
172
118
4
.32
.24
.16
.08
Ø
TRUE STRAIN
0.5
1.0
1.5
STRAIN RATE (1/s)

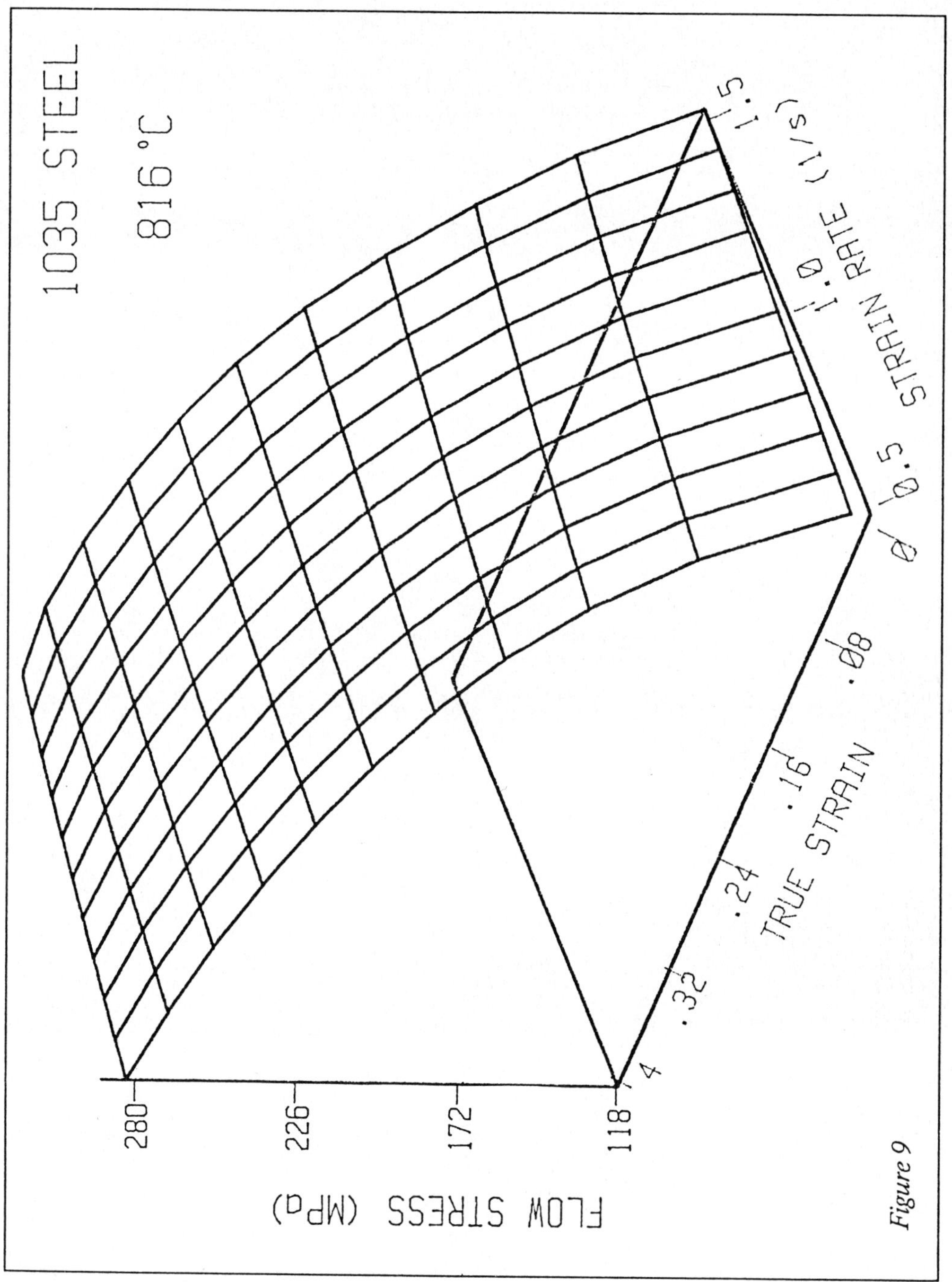

Figure 9

1035 STEEL
871 °C
FLOW STRESS (MPa)
227
178
129
80
.4
.32
.24
.16
.08
TRUE STRAIN
0
0.5
1.0
1.5
STRAIN RATE (1/s)

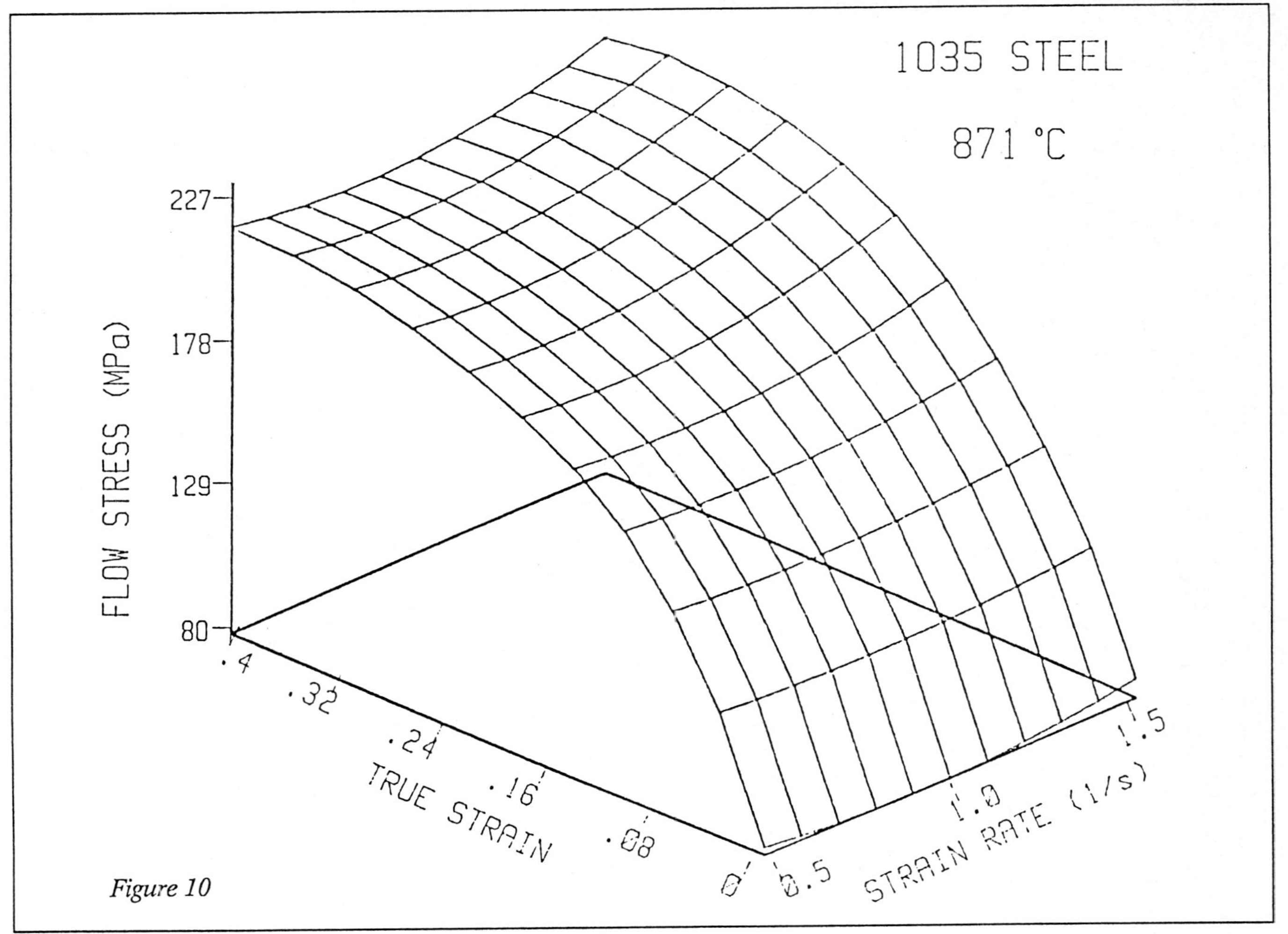

Figure 10

1035 STEEL
927 °C
FLOW STRESS (MPa)
157
123
89
55
.4
.32
.24
.16
.Ø8
Ø
TRUE STRAIN
Ø.5
1.Ø
1.5
STRAIN RATE (1/s)

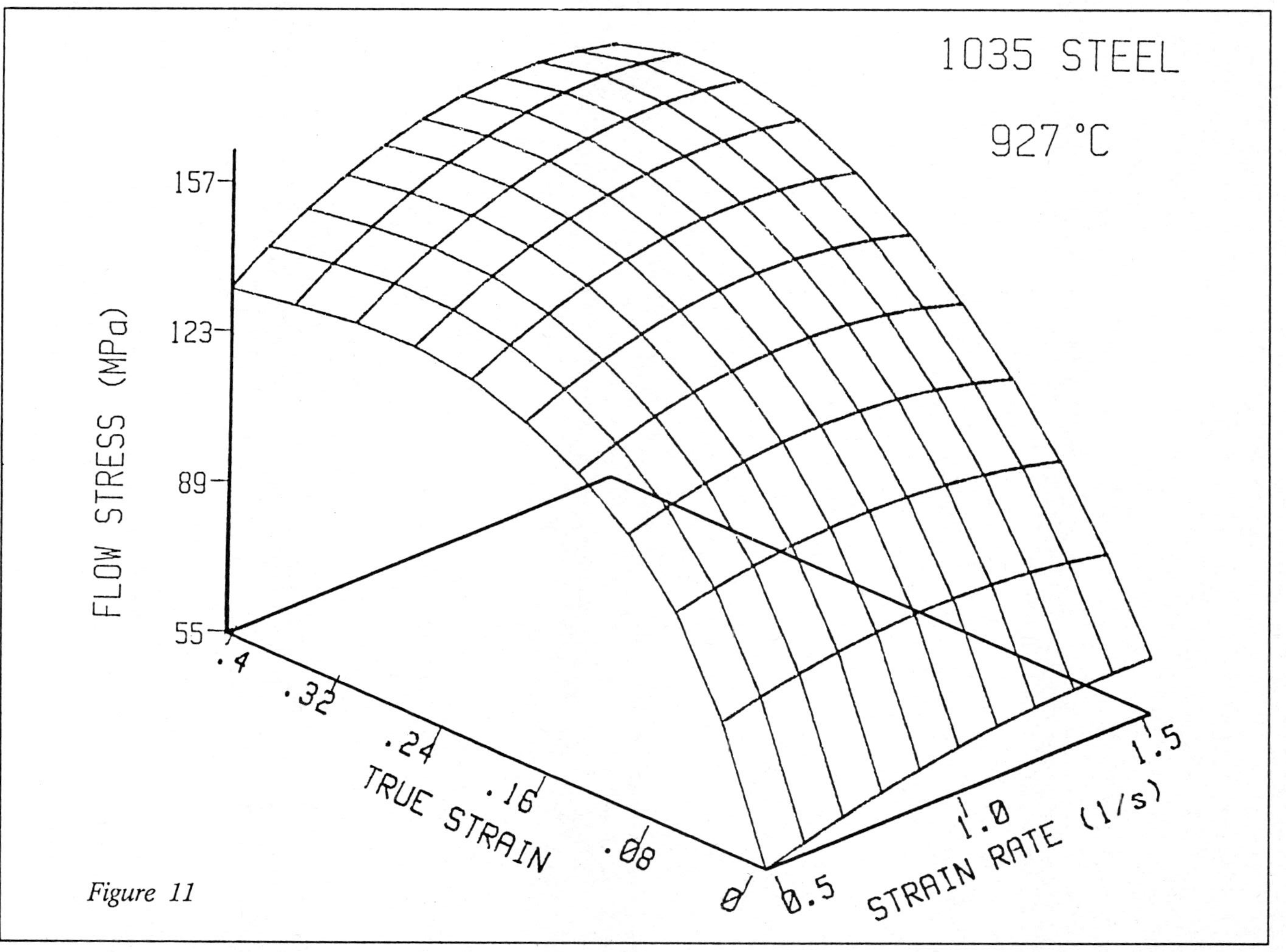

Figure 11

1035 STEEL
982°C
FLOW STRESS (MPa)
131
100
69
38
TRUE STRAIN
.4
.32
.24
.16
.08
0
STRAIN RATE (1/s)
0.5
1.0
1.5

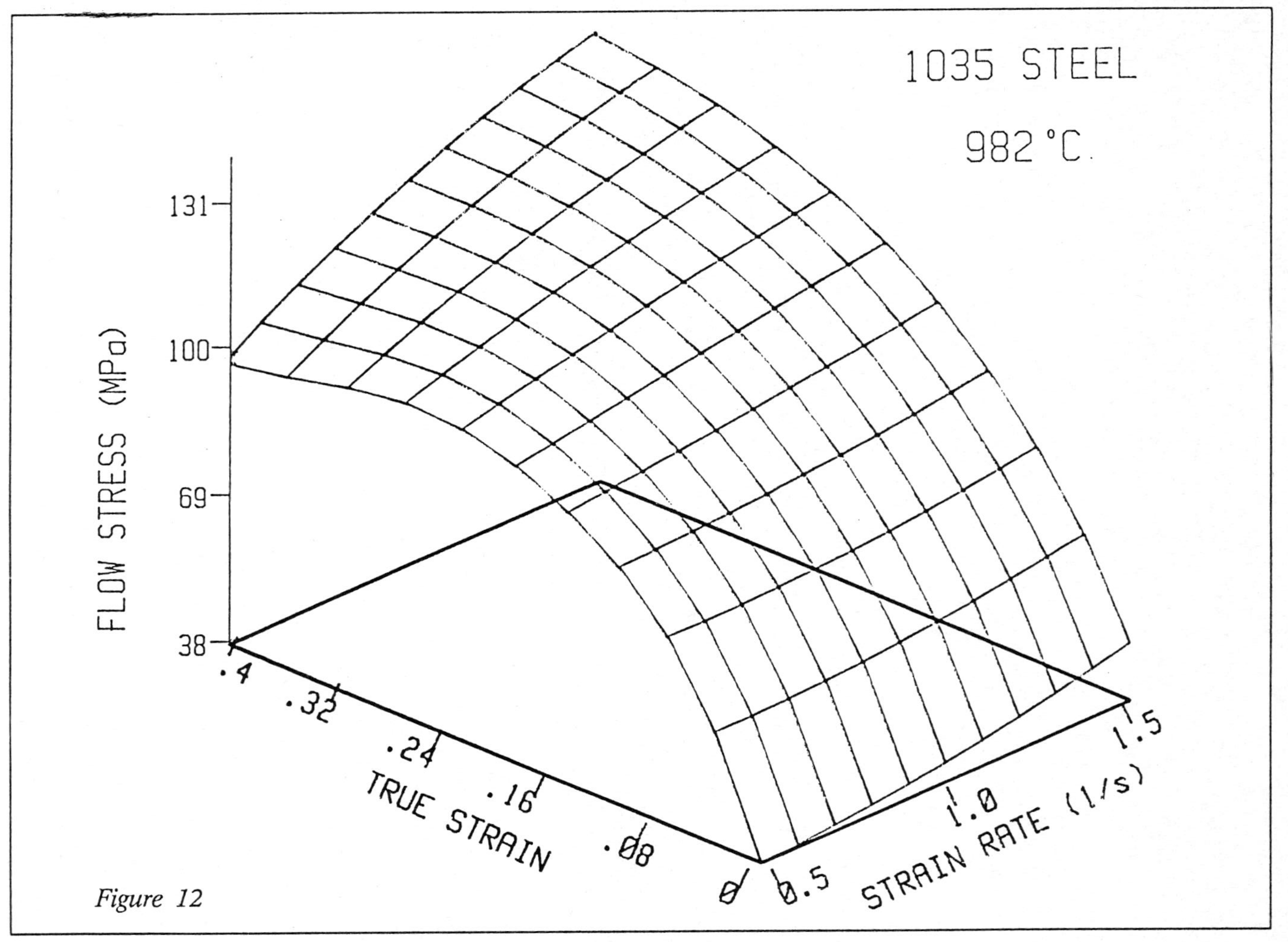

Figure 12

6. RESULTS

KAFTANOGLU and ONG [5] provided a three dimensional presentation of stress versus strain and strain-rate for temperatures 816°C, 982°C for AISI 1035 steel. Table 1 gives the material constants in the plastic range for Equation (14) (See also Eqn. (15.1)).

KADIOGLU tried 5 models to fit to AISI 1035 data. Typical computer outputs are given in Figures 3 to 7. KARAYAKA also applied Bezier curves and surfaces and B-spline curves and surfaces to KAFTANOGLU's AISI 1035 data. Typical 3D and 2D curves for AISI 1035 data representations are shown in Figures 8 to 12. Figure 8 represents 2D material behaviour at 1500°F (816°C) temperature. Others represent material behaviour at 816, 871, 927 and 982°C respectively. He obtained a Bezier fit with percent error ranging between 2.4 and 0.01% for 19 data points. Further he applied a numerical optimization technique to further improve the fit. For the same data set percent error ranged between 1.9% and 0.04% Figure 8 shows both fits before and after optimization.

YÜRÜR obtained experimental data for niobium steel at 13.3 l/s strain rate. He provided data for 20°C, 750°C, 850°C, 950°C with and without annealing.

Figures 13 and 14 gives true stress-true strain curves at room temperature in the rolling and transverse directions respectively.

High temperature behaviour of the same material before and after annealing is given in Figures 15, 16 and 17.

These models have been applied to Hot and Cold Rolling of Flat Strips [13], Roll-Forging [14] and Tube-Drawing [15].

7. CONCLUSIONS

Computer programs were developed to model the plastic material behaviour for ambient and high temperatures. Extensive experimental data were obtained for AISI 1035 and niobium steel. 3-dimensional stress, strain, strain-rate and 2-dimensional stress, strain curves were fitted using various techniques as being shown in this text. It has been observed that the curve fitting processes and the choice of constitutive equations are successful. This technique offers a promising design and analysis tool where it has been already applied to tube-drawing, roll-forging and hot and cold rolling of flat strip.

8. ACKNOWLEDGEMENTS

The author expresses his sincere thanks to the NATO authorities for their support of this project in the area of high temperature, high strain-rate behaviour of steel under contract No. 390/83. He is also thankful to Messrs. Yavuz Kadioglu, Cuneyt Yürür, and Metin Karayaka. Appreciation also goes to Drs. A. Nassirharand, K. Lange, Mr. A.S. Halal and How Rang Ong who collaborated in various aspects of this research by their technical and scientific contributions. Special thanks go to Dr. B. Kilkis who assisted in the preparation of this manuscript.

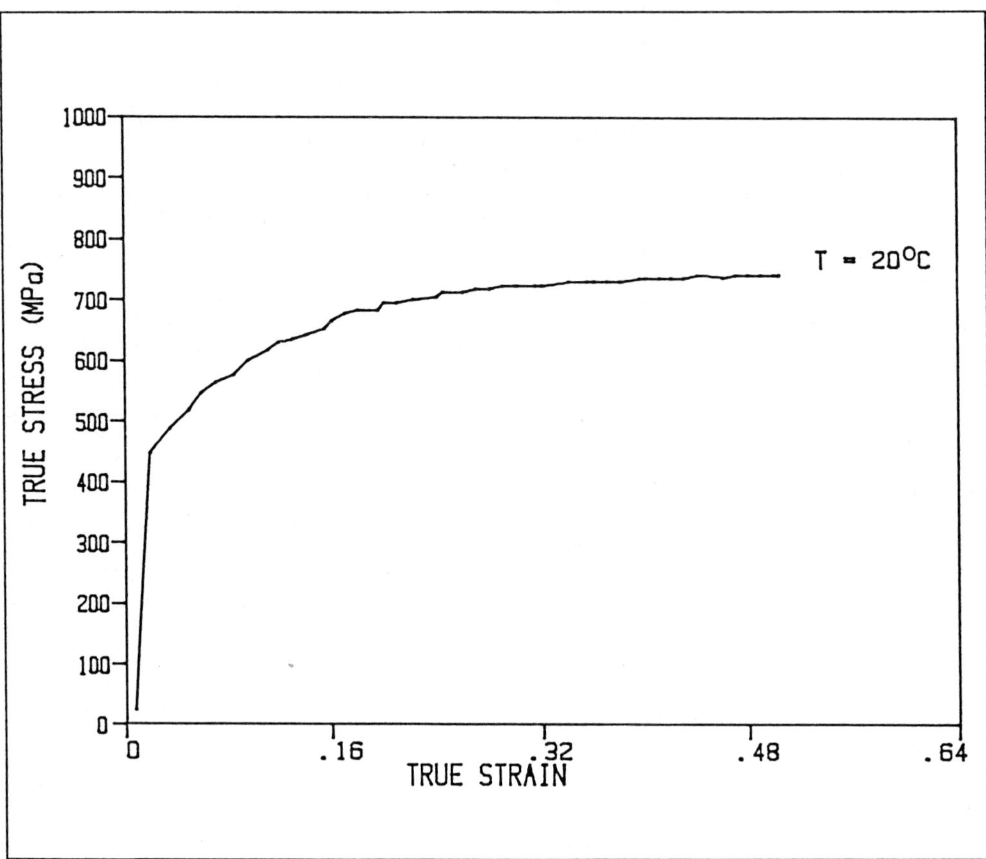

Figure 13: True Stress-True Strain Curve of Niobium Steel at Room Temperature (longitudinal rolling direction).

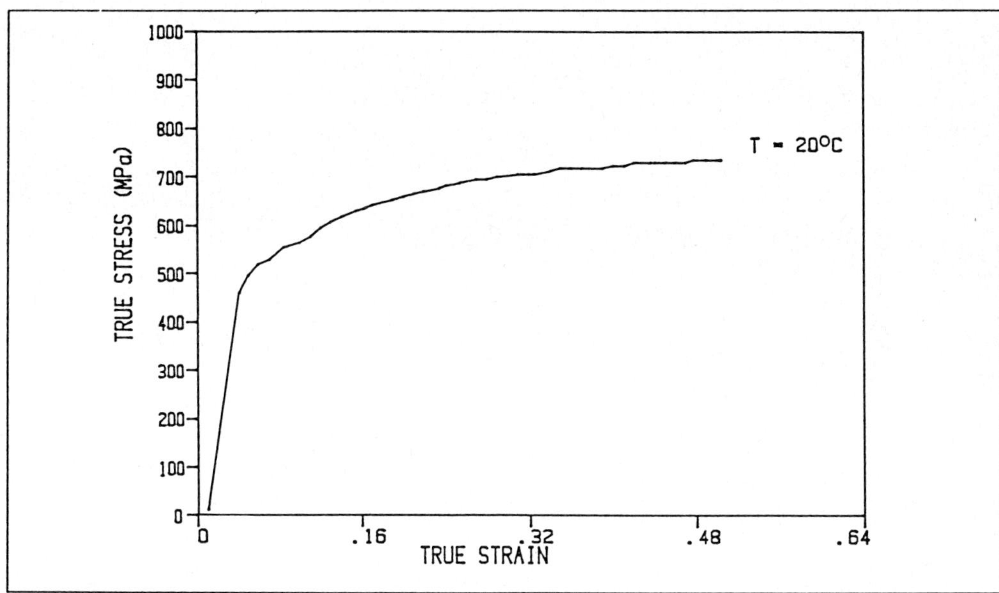

Figure 14: True Stress-True Strain Curve of Niobium Steel at Room Temperature (transverse r.d.) .

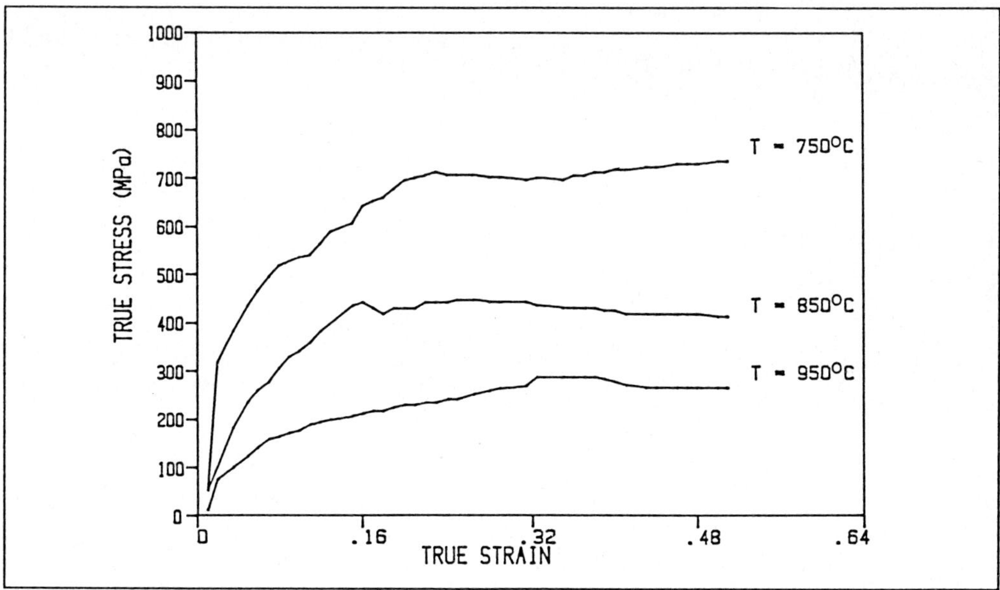

Figure 15: True Stress-True Strain Curve of Niobium steel at High Temperatures (without annealing) .

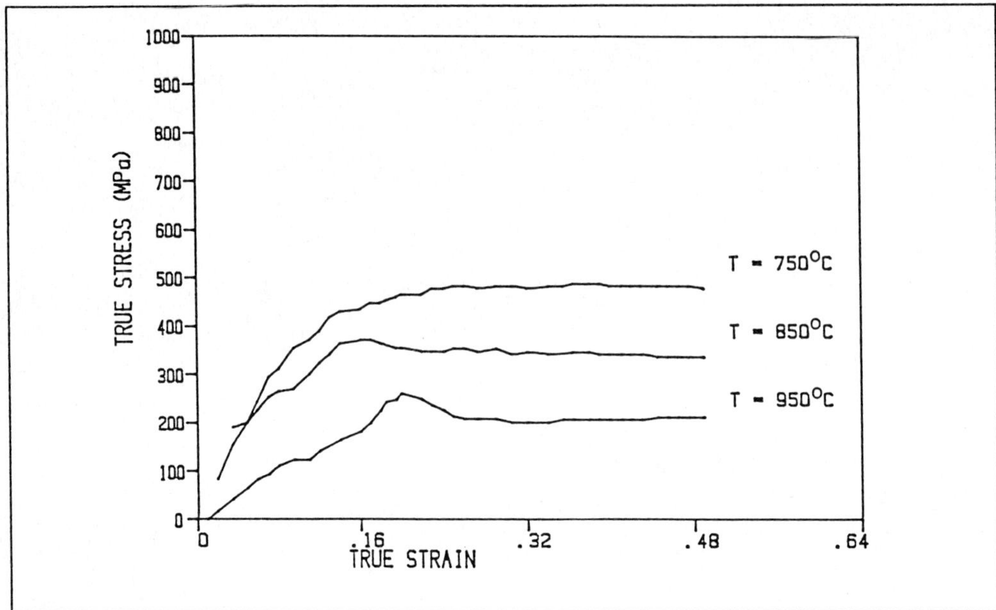

Figure 16: *True Stress-True Strain Curve of Niobium Steel at High Temperatures (after annealing).*

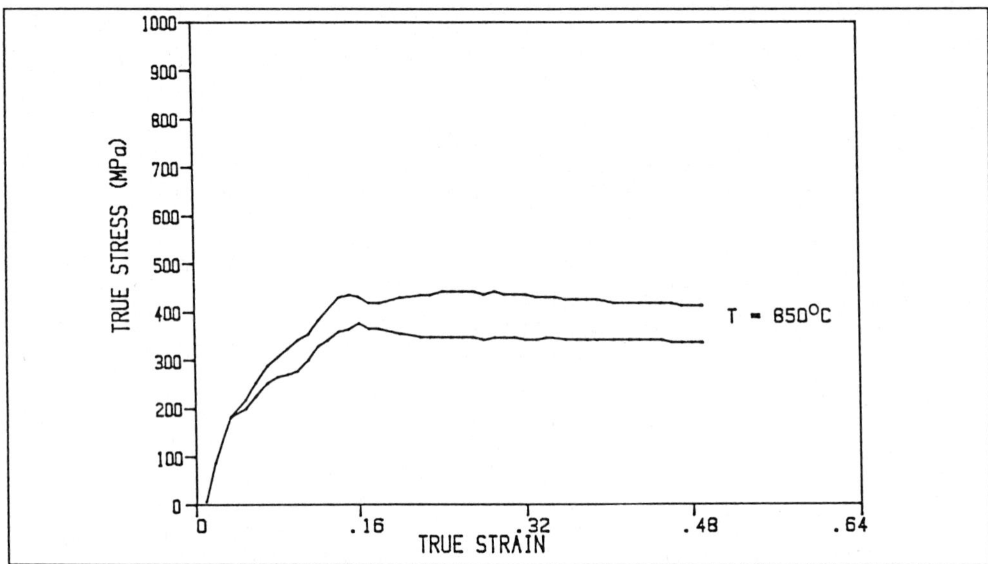

Figure 17: *The Effect of Annealing on the True Stress-True Strain Curves of Niobium Steel at T = 850°C.*

9. REFERENCES

1. Johnson, W. and Mellor, P.B., "Engineering Plasticity", Van Nostrand Reinhold, 1973.

2. Kaftanoglu, B. and Sivaci, K., J. of Eng. Mat. and Tech., 98, 1976, pp. 203-212.

3. Jonas, J.J., Sellars, C.M. and Tegart, W.J. McG., Metall. Rev. 14, 1969, p. 1.

4. McQueen, H.J. and Jonas, J.J., "Treatise on Materials Science and Technology", Academic Press, 1975.

5. Kaftanoglu, B. and Ong, H.R., "Modelling of Plastic Behaviour of AISI 1035 Steel at Elevated Temperatures", Journal de Physique, No. 8, T. 46, 1985, pp. C5 463-468.

6. Yürür, C., "Computer-Aided Testing of Metals at High Temperatures and Strain Rate", M.Sc. Thesis, Middle East Technical University, Ankara, 1986.

7. Kadioglu, Y., "Computer-Aided Development of Material Models", M.Sc. Thesis, Middle East Technical University, Ankara, 1986.

8. Karayaka, M., "Computer-Aided Modelling of Material Properties in Plastic Range", M.Sc. Thesis, Middle East Technical University, Ankara, 1987.

9. Hockett, J.E., "On Relating the Flow Stress of Aluminum to Strain, Strain Rate and Temperature", Trans. of the Metallurgical Society of AIME, 239, 1967, p. 969.

10. Blazynski, T.Z., "Metal Forming, Tool Profiles and Flow", Halsted Press, New York, 1986.

11. Marquardt, D.W., "An Algorithm for Least-squares Estimation of Nonlinear Parameters", J. Soc. Indust. Appl. Math.,11, 1963, p.431.

12. Rogers, D.F. and Adams, J.A., "Mathematical Elements for Computer Graphics", McGraw-Hill Book Co., New York, 1976.

13. Halal, A.S and Kaftanoglu, B., "Computer-Aided Modelling of Hot and Cold Rolling of Flat Strip",Proceedings of International Conference on Computers in Engineering (ASME), Las Vegas, 1984.

14. Kaftanoglu, B. and Nassirharand, A., "Computer Aided Roll-Forging", to be presented at 2nd Int. Conf. on Technology of Plasticity, 1987, Stuttgart, Fed. Rep. of Germany.

15. Kaftanoglu, B. and Lange, K., "Computer-Aided Modelling of Tube-Drawing", Annals of CIRP, 34/1/1985, pp. 249-252.

Chapter 5

Mapping Dynamic Material Behaviour

by J.M.Alexander

1. ABSTRACT

Perhaps the main problem facing the modern manufacturing engineer or materials scientist is that of adequately describing the mechanical and physical properties of the material he needs to process. In the past, a great deal of research work, both theoretical and experimental, has been directed towards determining material properties, generally in the form of true stress-true strain curves at various temperatures and strain-rates, and then endeavouring to find suitable empirical relationships to assist in cataloguing those experimental results to facilitate their use.

In the modern jargon, many attempts have been made in the past to establish a data bank or data base in a form which would enable prediction of the near-optimum processing conditions required for any given material to be made. The present position is that not much real progress has been made towards this goal. Quite often the basic data obtained from various supposedly reliable sources are at variance with one another and even if not, they give little information about undesirable or unstable conditions which might occur during processing.

The present paper is addressed towards trying to summarize the current state-of-the-art, to explain what is meant by terms such as constitutive equations, strain-rate sensitivity, processing efficiency, instability criteria, deformation and fracture mechanism maps, materials information systems, etc., and to predict what seems to be the most fruitful avenues for further research in this area, particularly in relation to the hot rolling of steels.

LIST OF SYMBOLS

a,m,n	material constants
s	entropy
A,B,D	material constants
G	dissipator content
J	dissipator co-content
Q	activation energy
P	power
R	universal gas constant
T	absolute temperature
T_H	homologous temperature
T_M	melting temperature
ε	true strain
$\dot{\varepsilon}$	strain rate
η	efficiency
σ	true stress

2. CONSTITUTIVE EQUATIONS

A discussion about constitutive equations has been given by ALEXANDER [1]. Approximate ranges of homologous temperature (T_H) and strain rate were investigated for many metal forming processes and are summarized in Table 1 below.

It may be concluded from this recent paper that one possible constitutive relation which might be useful for representing the dependence of the flow stress $\bar{\sigma}$ upon effective strain $\bar{\varepsilon}$, effective strain rate $\dot{\bar{\varepsilon}}$ and absolute temperature T is:

$$\bar{\sigma} = D\bar{\varepsilon}^{n_1}\dot{\bar{\varepsilon}}^{n_2}\exp(Q/RT) \qquad (1)$$

where Q is an activation energy,
R is the universal gas constant, and
D, n_1 and n_2 are constants.

TABLE 1

Typical ranges of T_H and strain rate

Process	Range of T_H	Approximate range of strain rate (s^{-1})
Casting	0.9-1.0	0.0-0.4
Hot forging	0.7	100-200
Hot extrusion	0.7	10-20
Cold forging and extrusion	0.16-0.32	10-100
Warm forging	0.55	10-100
Creep forming	0.63 +	10^{-2}-10^{-4}
Sheet metal forming	0.16-0.32	10-100
Powder forming	Not known	Not known
Deposition forming	0.9	200-1000
Machining, shearing	0.16-0.32	> 1000
Explosive forming	0.16-0.32	100-1000
Hot rolling	0.7	40-100
Cold rolling	0.16-0.32	100-1000

Since Equation 1 gives a zero value for $\bar{\sigma}$ if either $\bar{\varepsilon}$ or $\dot{\bar{\varepsilon}}$ is zero, which is certainly not the situation in practice, a more realistic equation might be:

$$\bar{\sigma} = \sigma_0(1 + B\bar{\varepsilon})^{n_1}(1 + D\dot{\bar{\varepsilon}})^{n_2}\exp(Q/RT) \qquad (1a)$$

where B is a constant.

A different approach, more geared to the dislocation mechanics of physical metallurgy, is that of ASHBY and FROST [2], who have developed "deformation mechanism maps" and "fracture mechanism maps", typically as depicted in Figure 1.

The strain rate of a plastically deforming crystalline solid is dependent upon a number of basic atomic processes such as: dislocation motion, diffusion, grain boundary sliding, twinning or a phase transformation. Such processes can combine to give at least 12 distinctive deformation mechanisms, e.g. yielding (or slip), power law creep, Nabarro-Herring creep, Coble creep, superplastic flow, etc.

Deformation mechanism maps for any polycrystalline material can be constructed in stress-temperature space, showing the area of dominance of each flow mechanism for each of which a rate equation exists, linking shear strain rate γ to shear stress τ, temperature T and to 'structure' (including all parameters describing its atomic structure, e.g. bonding, crystal class, defect structure, grain size, dislocation density and arrangement, solute or precipitate concentration, etc.). An example of such a map is shown in Fig. 1 on which typical equations have been shown, relating the normalized shear stress τ/μ, where μ = shear modulus, to homologous temperature. KELLY [3] has shown that τ cannot exceed τ_{ideal} ($\cong \mu/20$), shown by the line marked "ideal strength" in the figure, which is for pure nickel, but polycrystalline metals deform at stresses much less than this by dislocation glide and/or movement of vacancies. Examples of other maps and the methods of using them are described by ASHBY and FROST [2]. In some forming processes, where very large strain rate ranges may be encountered, the description of steady state behaviour given in Equation 1 is inadequate. At very high stress, an exponential stress dependence on strain rate is more appropriate and the power law and exponential variation can be combined as:

$$\bar{\dot{\varepsilon}} = A\{\sinh(a\bar{\sigma})\}^{m} \exp(-Q/RT) \qquad (2)$$

This relationship has been shown to give a good fit with experimental data for aluminium alloys over many order of magnitude change in strain rate.

However, the description is still of 'steady state' behaviour and there appears to be no good evidence that it can be used as an equation of state* relationship when stress or temperature change sharply. This problem has been approached in two ways.

One approach is to define an equation of state which contains an internal variable describing the structure of the material. If the variation of this internal variable with change in stress, temperature and time can be established, an adequate constitutive equation may result. Numerous attempts have been made to provide such a description but they all suffer from the disadvantage that large numbers of sophisticated tests have to be conducted to establish the internal variables. Thus not only con-

* An equation of state implies a fixed one-to-one relationship between dependent and independent variables, e.g.as for the equation relating the pressure,volume and absolute temperature of ideal gases in thermodynamics, viz: PV = RT.

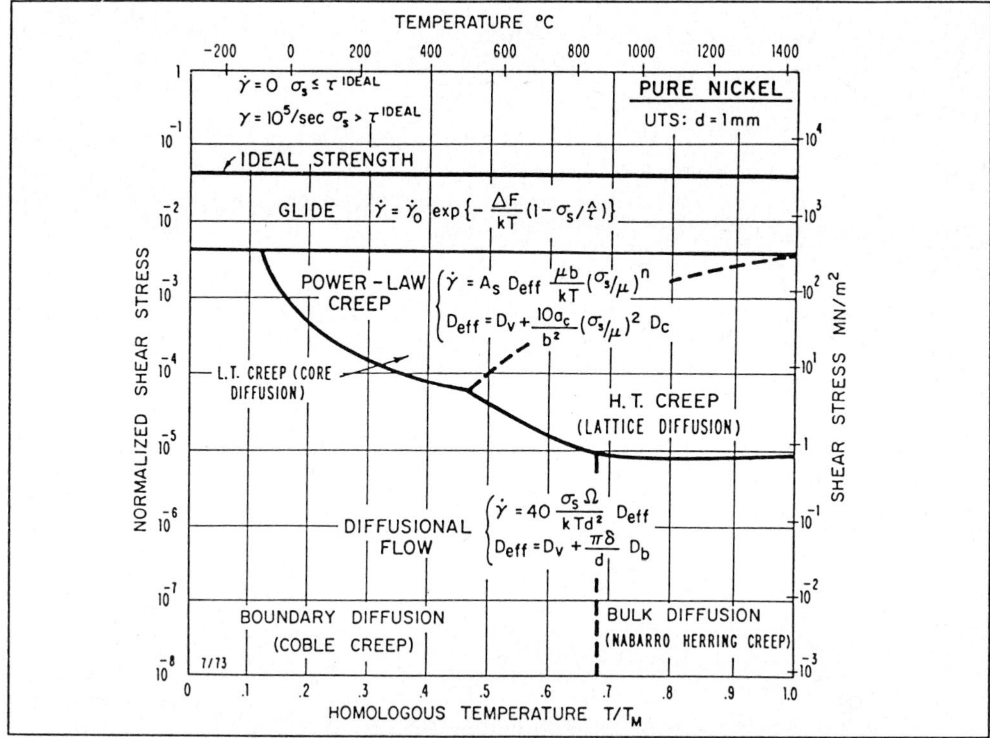

Figure 1: The fields in which a particular mechanism of flow is dominant. Boundaries are found by equating flow rates. (courtesy ASHBY et al.)

stant stress and constant strain rate tests are required but also complex relaxation and other stress and temperature cycling experiments.

Ashby's deformation and fracture mechanism maps have made an important contribution to the way of describing and illustrating both deformation and fracture phenomena. It is possible to delineate various regions for different deformation processes on a deformation mechanism map as illustrated in Figure 2 (for pure copper). It is worth noting here that ASHBY plots curves of normalized shear stress vs. homologous temperature, for various strain-rates, derived from dislocation mechanisms. The same approach can be used for fracture.

3. MODELLING OF DYNAMIC MATERIAL BEHAVIOUR IN HOT DEFORMATION

Another (similar) approach has been adopted by H. GEGEL and his co-re-

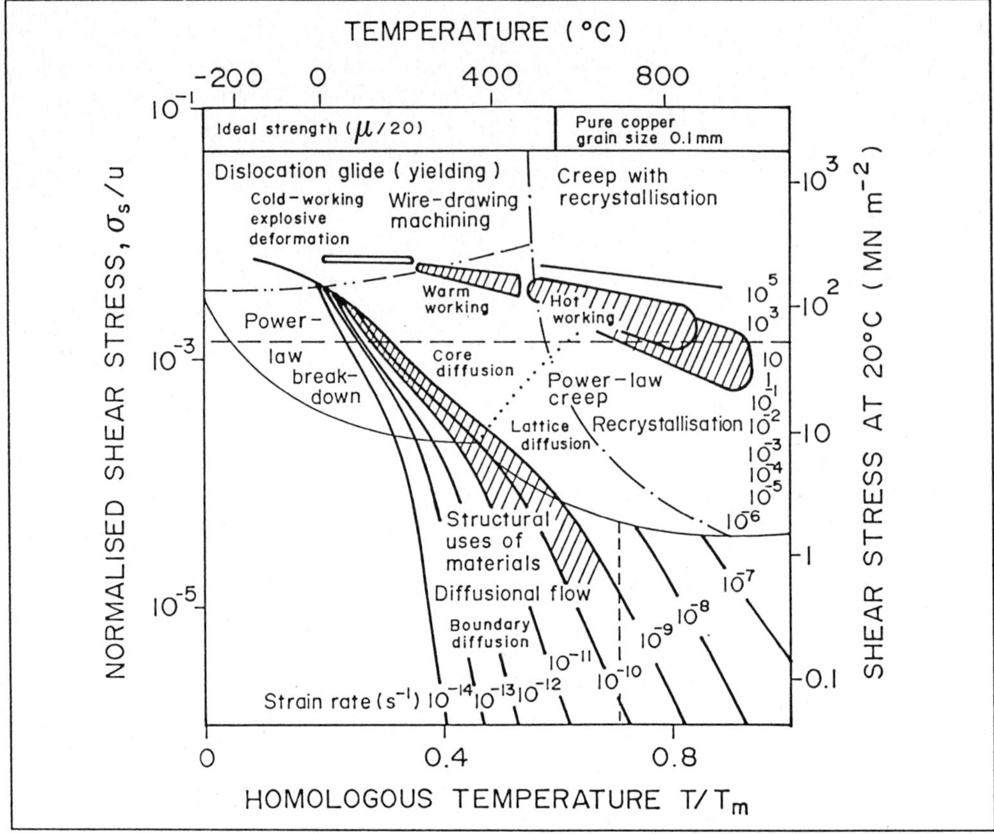

Figure 2: Forming operations on a deformation map for pure copper. (courtesy ASHBY et al.)

searchers, PRASAD et al. [4]. This approach, however, is based on thermodynamic considerations and does not rely on the microstructural behavioural description of the material. It is 'similar' only in so far as the final presentation is of areas in which deformation is possible without fracture or defects, and it is possible to identify optimum regions in which deformation is most efficient. This approach seems to be superior to that of ASHBY in that it is more general (not necessarily tied to metals) and is easier to understand. Attention is drawn towards the instantaneous power being dissipated, viz:

$$\bar{\sigma}\dot{\bar{\varepsilon}} = \int_0^{\dot{\bar{\varepsilon}}} \bar{\sigma}\, d\dot{\bar{\varepsilon}} + \int_0^{\bar{\sigma}} \dot{\bar{\varepsilon}}\, d\bar{\sigma} \tag{3}$$

or

$$P = G + J \tag{4}$$

where the power dissipated by plastic work is denoted by the area G under the curve in Fig. 3(a) and designated the "dissipator content" whilst the area J above the curve is termed the "dissipator co-content" and is related to the structural (metallurgical or non-metallic) mechanisms which occur dynamically to dissipate power.

From Equation 3 it follows that, at any given temperature and effective strain, the partitioning of power between J and G is given by:

$$\left(\frac{\partial J}{\partial G}\right)_{T,\bar{\varepsilon}} = \left(\frac{\partial \ln \bar{\sigma}}{\partial \ln \dot{\bar{\varepsilon}}}\right)_{T,\bar{\varepsilon}} \tag{5}$$

which is simply the strain rate sensitivity of a material conventionally evaluated as:

$$m = \frac{\partial(\ln \bar{\sigma})}{\partial(\ln \dot{\bar{\varepsilon}})_{T,\bar{\varepsilon}}} \cong \frac{\Delta(\log \bar{\sigma})}{\Delta(\log \dot{\bar{\varepsilon}})_{T,\bar{\varepsilon}}} \tag{6}$$

(Note: m = n2 of the discussion relating to Equation 1.)

Determining the values of m as a function of effective strain rate for a material and assuming that the exponential equation:

$$\bar{\sigma} = A\dot{\bar{\varepsilon}}^{m} \tag{7}$$

is valid for each value of m, the values of J are then calculated as a function of effective strain rate at a given T and $\bar{\varepsilon}$. A piece-wise quadratic fit is used to determine the various values of m over the range of strain rate used in the tests.

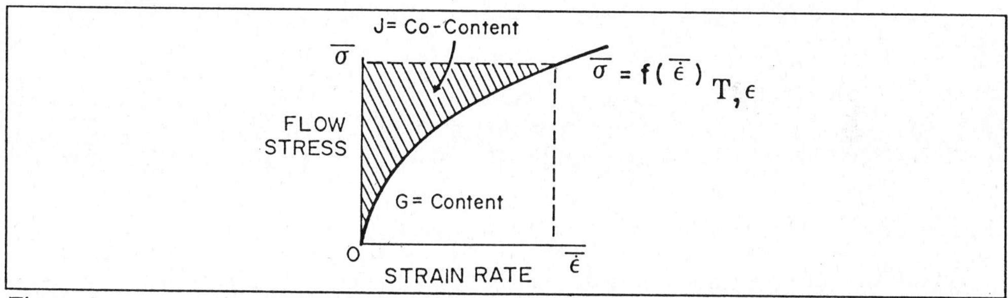

Figure 3: (a) Schematic representation of G content and J co-content for work piece having a constitutive equation represented by curve $\bar{\sigma}=f(\dot{\bar{\varepsilon}})$. Total power of dissipation is given by the rectangle [4].

Thus the power dissipated in changing the material structure is:

$$J = \int_0^{\bar{\sigma}} \bar{\dot{\varepsilon}} d\bar{\sigma} = \int_0^{\bar{\sigma}} (\bar{\sigma}/A)^{1/m} d\bar{\sigma} = \frac{1}{A^{1/m}} \frac{\bar{\sigma}^{(1/m+1)}}{(1/m+1)} = \frac{\bar{\dot{\varepsilon}}\bar{\sigma}m}{m+1} \qquad (8)$$

GEGEL and his colleagues show elsewhere that m must lie between 0 and 1 if the material deformation is not to become unstable. For values of $m < 1$ the curve of $\bar{\sigma}$ versus $\bar{\dot{\varepsilon}}$ appears similar to that shown in Fig. 3(a). When $m = 1$, the curve is a straight line giving the maximum dissipator co-content (of area $J_{max} = \bar{\dot{\varepsilon}}\bar{\sigma}/2$), as shown in Fig. 3(b).

Hence, it seems sensible to define an efficiency of deformation in terms of a non-dimensional parameter, in line with other stability criteria, namely

$$\eta = \frac{J}{J_{max}} = \frac{2m}{m+1} \qquad (9)$$

These researchers then proceeded to determine curves of σ and ε for gas turbine disc materials such as Ti-6242 at various temperatures and strains, by cross plotting and interpolation of data obtained from conventional stress strain compression testing up to high strain values. A three dimensional plot of efficiency η vs. temperature T and logarithmic strain-rate $\ln\bar{\dot{\varepsilon}}$ can then be constructed and two-dimensional maps showing η contours derived for various values of effective strain $\bar{\varepsilon}$, as shown typically in Figs. 4 and 5.

It is also possible to delineate regions on these maps where fracture or defects are most likely to occur as shown in Fig. 6 and this is research currently being undertaken by the group under GEGEL's leadership. For example, the parameter $\sigma_m/\bar{\sigma}$ (where σ_m = the mean or hydrostatic stress) can be determined everywhere throughout the volume of material being deformed (e.g. by finite element methods) and it must be ensured that it does not exceed algebraically a certain limiting (compressive) value, typically between -2/3 for extrusion at the exit to avoid internal cracking (centre burst). This parameter, taken in conjunction with dynamic recrystallization, phase transformations, spheroidization of acicular structures, precipitation mechanisms, edge cracking and local kinking of β platelets in Ti-6242 can all contribute to prediction of zones on these maps which should be avoided. It is very interesting to observe that the predicted optimum forging conditions of both $\alpha\beta$ and β preform microstructures of Ti-6242 are 927°C and 10^{-3} s^{-1}, corresponding almost exactly with the creep-forming conditions which have been arrived at in practice by laborious and expensive trial and error methods.

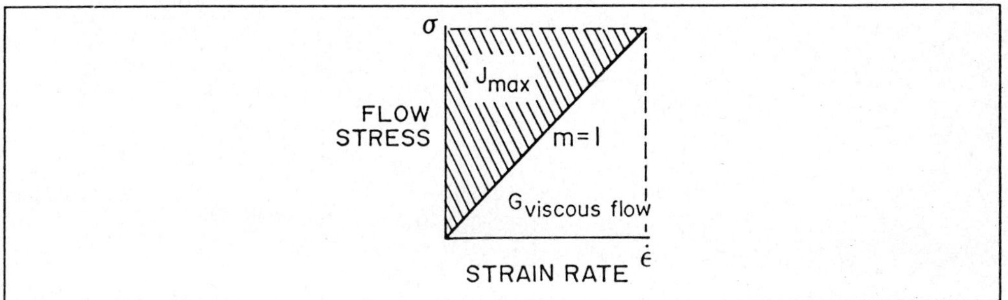

Figure 3 (b): Schematic representation showing J_{max} which occurs when the strain-rate sensitivity (m) of the material is equal to one. [4]

More recently, by applying Liapunov function stability criteria (well-known in control theory, e.g. as described by SCHULZ and MELSA [5]) to material processing variables such as m and introducing an analogous entropy coefficient s related to stress $\bar{\sigma}$ and absolute temperature T by the equation:

$$s = \frac{1}{T} \frac{\partial(\log \bar{\sigma})}{\partial(1/T)} \Big|_{\bar{\varepsilon}, \dot{\bar{\varepsilon}}} \qquad (10)$$

GEGEL and his colleagues have pointed out that:

$$\dot{s} = \dot{s}_{sys}/\dot{s}_{app} \qquad (11)$$

where $\dot{s}_{sys}$ = the rate of entropy production in the system and $\dot{s}_{app}$ = the rate of entropy input into the system.

The analogy between m and s can best be seen by comparing Equation 6, viz:

$$m = \frac{\partial(\ln \bar{\sigma})}{\partial(\ln \dot{\bar{\varepsilon}})_{T, \bar{\varepsilon}}} \cong \frac{\Delta(\log \bar{\sigma})}{\Delta(\log \dot{\bar{\varepsilon}})_{T, \bar{\varepsilon}}} \qquad (6)$$

with Equation 10.

Using the Liapunov function stability criteria then leads to the condition that, in stable regions, the two second order partial differential coefficients of m and s with respect to log $\dot{\bar{\varepsilon}}$ should be negative, viz:

$$\frac{\partial m}{\partial(\log \dot{\bar{\varepsilon}})} < 0 \; ; \quad \frac{\partial s}{\partial(\log \dot{\bar{\varepsilon}})} < 0 \qquad (12)$$

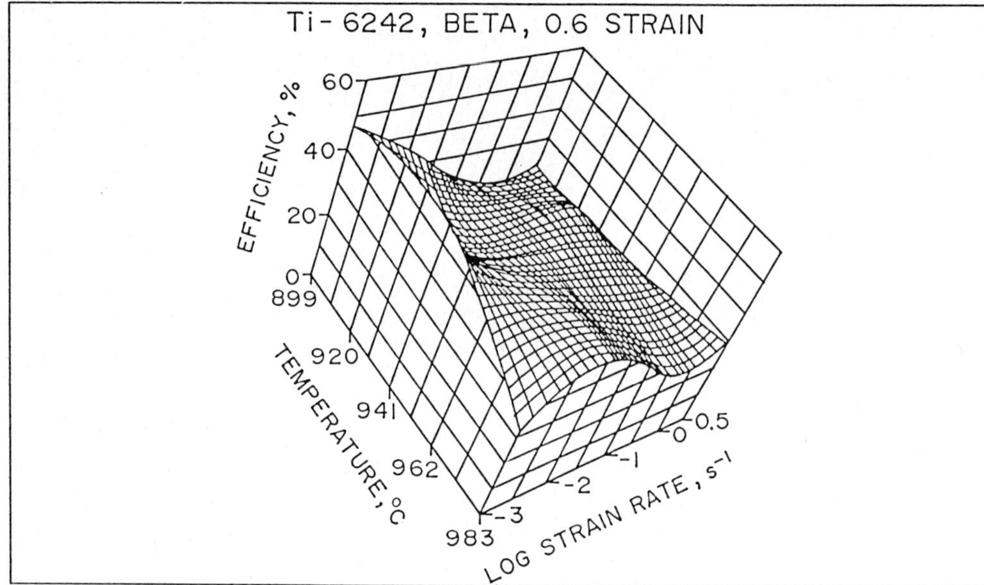

Figure 4: *Three-dimensional plot showing variation of efficiency of dissipation with temperature and strain-rate for Ti-6242 preform at 0.6 strain. [4]*

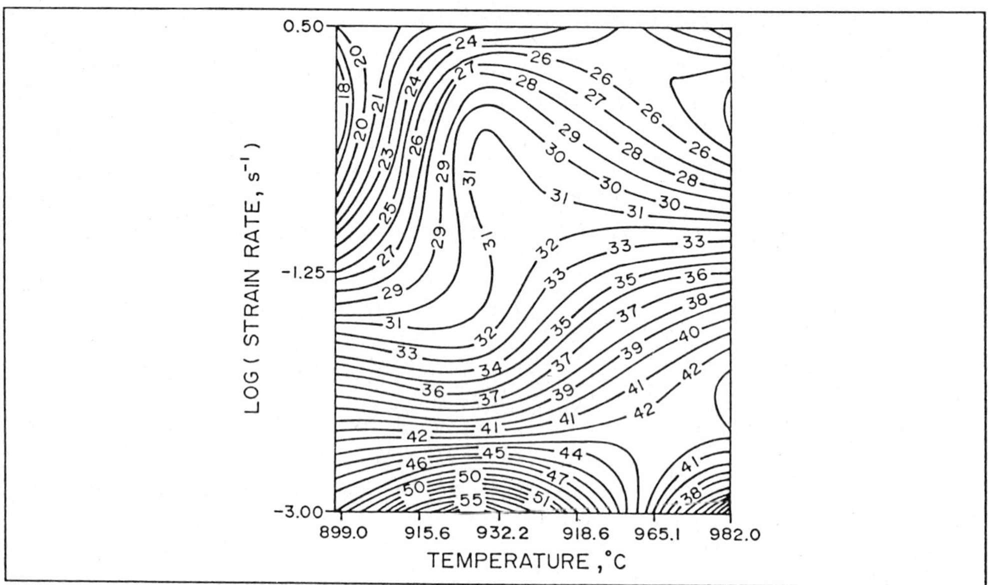

Figure 5: *Map showing constant efficiency contours in strain rate-temperature frame for Ti-6242 β-preform at 0.6 strain. [4]*

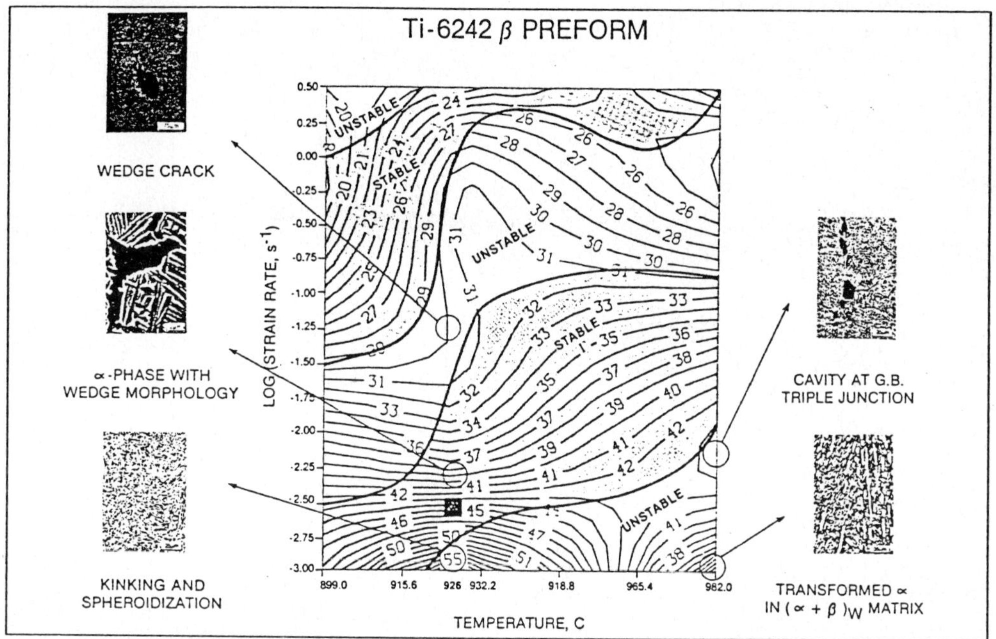

Figure 6: Processing map for Ti-6242 β microstructure with stable regions identified. [10]

When the rate of change of the slope of a function is negative, this implies that the function itself is a maximum. For the material parameters m and s, therefore, the condition for stability is that they should both be as large as possible, within the obvious limitations that neither can exceed unity. (Because m = 1 corresponds with Newtonian fluid viscosity and s = 1 corresponds with the rate of entropy production in the system being equal to the rate of entropy being put into the system).

These two conditions are used in the development of processing maps of the type shown in Figures 6 and 7, as reported by GOPINATH [6].

4. MODELLING OF THE HOT FLAT ROLLING OF STEEL: SUMMARY AND CONCLUSIONS

It seems that the mapping techniques being developed by GEGEL and his group offer the most useful approach for developing better mathematical modelling of material behaviour. Data bases should be set up for several materials of interest, using the same techniques which have given such excellent predictions of actual behaviour for the materials used by GEGEL.

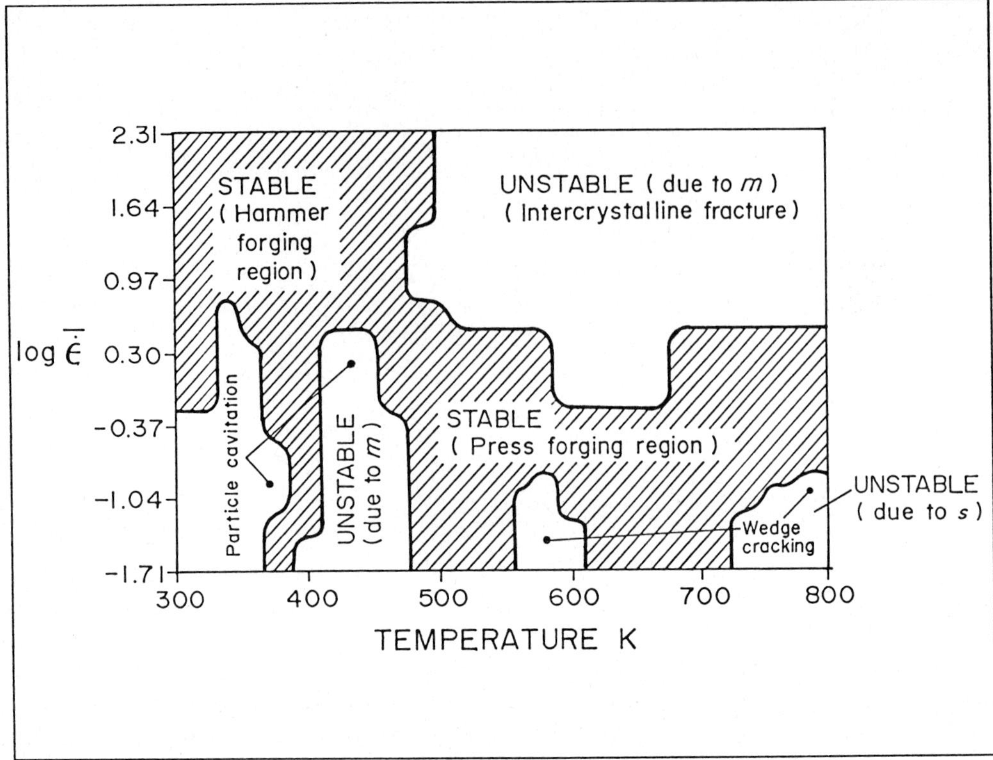

Figure 7: Stability map showing safe processing regions for Al-5Si alloy, at $\bar{\varepsilon}$ = 0.6. (courtesy of GOPINATH [6])

At first sight, many researchers find difficulty in accepting the concept of dissipator content and co-content. Also, it has been observed that variables such as T and ε are being treated as state variables (much as pressure, volume and temperature in gas thermodynamics), whereas they must be dependent on the history of their development. For example, if a metal is taken to a certain value of temperature and total strain without passing through any phase transformation it will then have a different stress/strain-rate relationship than if it <u>had</u> have passed through a phase change.

However, the method manifestly works. It is not actually necessary to use the concept of efficiency, viz. $\eta = J/J_{max}$. However, it is easier to correlate the parameter η with the microstructures evaluated over various regions of the processing maps and thereby arrive at more meaningful conclusions for understanding the intrinsic workability of materials. Similar results could be obtained simply by maximizing the m-value of Equation 6, (up to its maximum stable value of unity), i.e. by developing three-dimensional plots and mappings with m as the ordinate rather than η = 2m/(m + 1), since:

$$\eta m + \eta = 2m, \quad m(2 - \eta) = \eta, \text{ and } m = \eta/(2 - \eta) \tag{13}$$

There is, in fact, almost a linear relationship between m and η for values of m and η between 0 and 1 as shown in Figure 8.

Since a value of m = 1 corresponds with Newtonian viscosity, i.e. the viscosity possessed by most simple liquids, this is obviously the most efficient way to deform a material (by making it approximate to the liquid state). The higher the value of m, the easier (more efficient) is the process of deformation going to be (e.g. metals becoming "super-plastic"). However, it is not possible so easily to correlate the value of m with observed or predicted microstructures as is the efficiency η.

It may be concluded that the best approach to constitutive equations is that of developing deformation and fracture mechanism maps, on the lines proposed by PRASAD et al. [4]. The "deformation efficiency" concept is attractive and has demonstrably led to a better understanding of the important parameters to be developed to give a meaningful data base for any given material of interest. The Liapunov function stability criteria seem to provide powerful tools for delineating stable and unstable regions on these mappings of deformation behaviour. Apparently, these regions so discovered do actually correspond to material instabilities well-known to materials scientists dealing with both metals and non-metals. One remaining problem, however is that of adequately accounting for the history of the deformation.

Modelling of dynamic material behaviour in hot deformation can most easily be included by using this mapping approach. The data base so developed could then be used to provide data as a function of temperature, strain, strain rate and time (to include the history of deformation) which can then be fed into comprehensive computer solutions for rolling of the type already developed by ALEXANDER [7], VENTER and ABD-RABBO [8] and HALAL and KAFTANOGLU [9].

However, these computer solutions are all of the slab type and do not include either temperature or strain-rate effects. Much work has been done in industry on this problem, but unfortunately it has not been published widely.

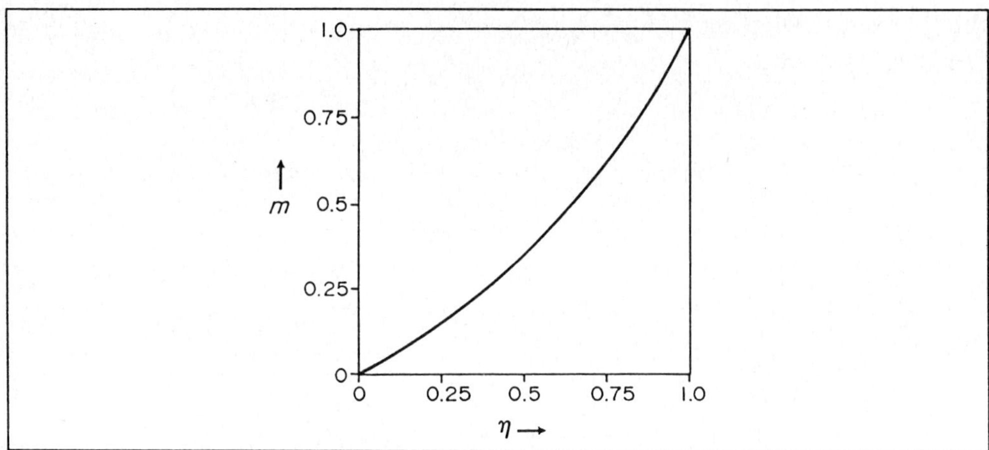

Figure 8: Variation of m with η for equation 13.

5. REFERENCES

1. Alexander, J.M., "On Problems of Plastic Flow of Metals", Plasticity Today, (Ed. H. Sawczuk), W. Olzak Memorial Volume, Int. Centre for Mech. Sci., Udine, Italy, June 1983. To be published by Elsevier, App. Sci. Publishers Ltd., 1987.

2. Ashby, M.F. and Frost, H.J., "The Kinetics of Inelastic Deformation above 0°K", Constitutive Equations in Plasticity (Ed. A.S. Argon), M.I.T. Press, 1975, 117.

3. Kelly, A., "Strong Solids", Clarendon Press, Oxford, 1966.

4. Prasad, Y.V.R.K., Gegel, H.L., Doraivelu, S.M., Malas, J.C., Morgan J.T., Lark, K.A. and Barker, D.R., "Modelling of Dynamic Material Behaviour in Hot Deformation: Forging of Ti 6242". Met. Trans. A. 15A, 1984, 1883.

5. Schultz, D.G. and Melsa, J.L., State Functions and Linear Control Systems, McGraw-Hill, 1967, 155-195.

6. Gopinath, S., Automation of the Data Analysis System Used in Process Modelling Applications, M.S. Thesis, Ohio University, Athens, Ohio 1986.

7. Alexander, J.M., "Micro Computer Programs for Rolling and Extrolling", Metal Forming and Impact Mechanics (Ed. S.R. Reid), William Johnson Commemorative Volume, Pergamon Press, 1986, 91.

8. Venter, R.D. and Abd-Rabbo, A.A., "Modelling of the Rolling Process - I and II", Int. J. Mech. Sci., 1980, 22, 83-92 and 93-98.

9. Halal, A.S., and Kaftanoglu, B., "Computer-Aided Modelling of Hot and Cold Rolling of Flat Strip", A.S.M.E. Int. Computer in Eng. Conf. Las Vegas, Nevada, 1984.

10. Gegel, H.L., Malas, J.C., Doriavelu, S.M., Alexander, J.M., and Gunasekera, J.S., "Materials Modelling and Intrinsic Workability for Simulation of Bulk Deformation, 2nd Int. Conf. on the Technology of Plasticity, 1987, Stuttgart, FRG.

Appendix

Flow Curves of Microalloyed Steels

The results of experiments conducted at the Institut für Umformtechnik, Stuttgart, Federal Republic of Germany and The University of New Brunswick, Fredericton, Canada

1. INTRODUCTION

Computer integrated metal forming processes make use of, among other things, predictive-adaptive mathematical models that calculate parameters and control events. An essential portion of these models is an adequate description of the metal's response to forming. This response is a direct function of the thermal-mechanical treatment, that is, of the prior heating schedule, number of passes, strains/pass, strain rates, temperatures, delay times between passes and cooling rates.

Numerous publications, dealing with the material's resistance to deformation, may be found in the technical literature. Flow curves obtained in uniaxial tension, torsion and compression have been presented. As well, empirical relations for the flow strength as a function of various parameters have been developed. Since the metal of concern in this project is steel, it is of some interest to compare the strengths of various steels as predicted by a few of the existing formulae.

For low carbon steels, useful empirical formulae have been presented by

SHIDA [1], giving the flow strength in terms of strain, strain rate, temperature and carbon content. Strength and strain rate hardening coefficients for high temperature behaviour of several steels have been compiled by ALTAN and BOULGER [2]. Constitutive relations have also been given by GITTINS et al. [3], HAJDUK [4], GELEJI [5], EKELUND [6] and CORNFIELD and JOHNSON [7]. Comparison of their predictions, however, reveals that the information they present is far from convincing. For example, considering an AISI 1015 steel subjected to a true strain of 50%, at a constant strain rate of 20 s^{-1} and at 1200°C, SHIDA [1] predicts the flow strength to be 97 MPa, ALTAN and BOULGER [2] predict 78 MPa, CORNFIELD and JOHNSON [7] give 7 MPa while EKELUND [6] prescribed 34 MPa. For other process parameters the data calculated are similarly inconsistent.

The state of affairs is not different when the flow curves for microalloyed steels are considered - see, for example, References [8], [9] and [10]. It is observed that σ_p and ε_p, the peak strength and strain, depend very strongly on the prior heat treatment as well as on the mode of testing. Comparing the results of References [8], [9] and [10], conducted in torsion, tension and compression, respectively, one notices significant variations, not always consistent with expectations. The peak strength of 0.05% Nb bearing steel, at 950°C and at a strain rate of 0.1 sec^{-1} is given as 121 MPa by [8]; for a steel containing 0.038% Nb and about 50% more Mn, at ε = 0.005 sec^{-1}, Reference [9]'s peak strength is approximately 60 MPA, measured in tension. In compression testing of 0.02% Nb steel, the peak strength, at ε = 0.2 sec^{-1} is found to be 65 MPa by Reference [10].

The differences in prior thermal treatment are, of course, partially responsible for these variations. These, however, do not explain the differences in the calculated values of the activation energy and rate-hardening exponents. Reference [8] gives 434 kJ/mole and 8.6 respectively; Reference [10] gives 433 kJ/mole and 7.5 while Reference [9] calculates 327 kJ/mole and 4.6.

It is these discrepancies that guided the present authors to define the objectives of the experiments for this project. A program to develop flow strength data for use in metal forming analyses was initiated. Constant strain rate compression was decided to be used and the behaviour of two microalloyed steels when subjected to single and multistage compression is presented.

The experiments were carried in the laboratories of the Institut für Umformtechnik, Universität Stuttgart, under the direction of Prof. K. Lange and in the Manufacturing Processes Laboratory of the University of New Brunswick, under the supervision of Prof. J.G. Lenard.

2. FLOW CURVES OF STEEL #1

The steel to be tested was hot strip of 12.6 mm thickness. The chemical analysis and some mechanical properties are given in Table A1 below. Specimens for upsetting tests were machined out of the material as follows:

height: 12.0 mm
diameter: 8.0 mm

direction of specimen axis:

type (1): in rolling direction (standard)
type (2): in transverse direction
type (3): in normal direction

Upsetting tests were carried out at room temperature for all three types of specimens, and for type (1) also at 500°C and 700°C.

For the temperatures 500°C and 700°C the effect of strain rate on flow stress was studied by varying the speed of movement of the upsetting dies through the number of strokes of the excenter press. By this, two different values of $\dot{\varepsilon}_{av}$, the strain rate averaged over the duration of deformation, were obtained.

The test results are shown in Figures A1 and A2.

Discussion of test results

From Figure A1 it can be seen that there was almost no effect of directionality on the flow curve. Figure A2 demonstrates the effects of temperature and strain rate for type (1) specimens. For not too low strains, the flow curves can be shown to fulfill the analytic equation $\sigma_f = C\varepsilon^n$. The values of the constants C and n are given in Table A2 below.

Assuming that the flow stress is proportional to $\dot{\varepsilon}^m$ the strain rate sensitivity index m is obtained from the test results as follows:

m = + 0.11 ± 0.04 for 500°C
m = + 0.18 ± 0.04 for 700°C

The microstructure of the X70 steel is shown in Figure A3.

THYSSEN STAHL AKTIENGESELLSCHAFT
Forschung
Warmgewalzte Produkte
Stahlentwicklung

Ergebnis der Werkstoffuntersuchung

Stahlsorte: X 70

Walzerzeugnis: Warmband

Dicke (mm): 12,6

Anlieferungszustand: Thermomechanisch gewalzt

Wärmebehandlung: keine

Chemische Zusammensetzung (%)

Schmelze	Warmband-Nr.	Wb.-dicke (mm)		C 10^{-2}	Si 10^{-2}	Mn 10^{-2}	P 10^{-3}	S 10^{-3}	N 10^{-3}	Al 10^{-3}	Cr 10^{-2}	Cu 10^{-2}	Ni 10^{-2}	Nb 10^{-3}	V 10^{-2}	Zr 10^{-2}
178 944	255 378	12,6		12	33	142	16	6	6	20	4	3	3	46	8	-

Festigkeitseigenschaften

	Wärmebehandlung	Probenlage	Probenform	Prüftemperatur °C	Streckgrenze R_{eL} N/mm²	Streckgrenze $R_{p\,0{,}2}$ N/mm²	Zugfestigkeit R_m N/mm²	Reißfestigkeit N/mm²	Verhältnis $\frac{R}{R_m} \times 100$ %	Bruchdehnung A_5 %	Brucheinschnürung Z %	Warmstreckgrenze Prüftemp. °C	Warmstreckgrenze R N/mm²
255 378	keine	längs	Flachzug	RT	514	-	651		79	24	-		
		quer			543	-	664		82	23	-		
"	"	diagonal ∢ 30°	"	"	540	-	659		82	26	-		

Zähigkeitseigenschaften

Wb.-Nr.	Wärmebehandlung	Probenlage	Probenform	Kerbschlagarbeit (J) -100°C	-80°C	-60°C	-40°C	-20°C	0°C	+20°C	+40°C	+60°C	+80°C	+100°C	+120°C
255 378	keine	lgs	ISO-V	6	45	88	115	141	170	159					
		quer				20	37	49	55	52	55	57			
		diagonal ∢ 30°			13	34	58	68	70	77	74				

PA 4999

Nr. 1235 c/50

Table A1: Mechanical Properties of "X70" Steel

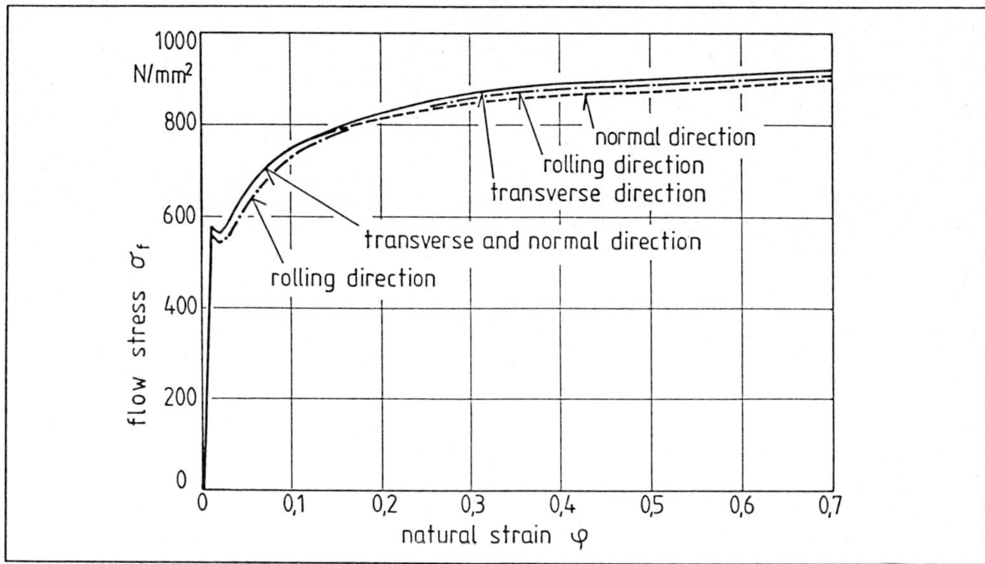

Figure A1: Flow curves at room temperature.

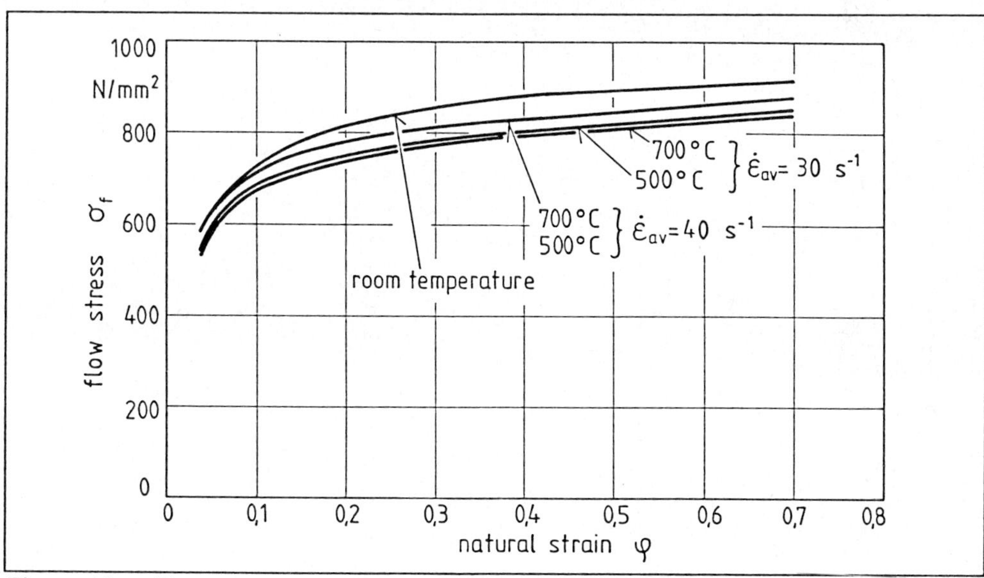

Figure A2: Flow curves at 500°C and 700°C.

3. FLOW CURVES OF STEEL #2

The material was obtained from the Republic Steel Co., Ltd, U.S.A. in the form of hot rolled strips.

Chemical Composition

C	Mn	Si	P	Al	Nb	Fe
0.09	0.90	0.02	<0.008	0.04	0.02	rest

Sample Preparation

Specimens of 4.7 mm diameter and 5.7 mm height were machined in direction 1 and samples of 5.2 mm diameter and 7.8 mm height were cut in direction 3 - see Figure A4. The ends of all samples were machined flat.

Lubrication

Before each test the specimens were dipped in a glass powder - alcohol suspension to reduce interfacial friction. Small amount of barrelling was observed

Temperature °C	Average Strain Rate s^{-1}	Coefficient C N/mm^2	Strain Hardening Coeff. n
20	0.03	970 $\pm$20	0.08 $\pm$0.01 for $\varepsilon \geq 0.2$
500	33	930 $\pm$20	
500	43	960 $\pm$20	
700	33	890 $\pm$20	0.12 $\pm$0.01 for $\varepsilon \geq 0.1$
700	43	930 $\pm$20	

Table A2: Values of coefficients C and n in the equation $\sigma f(\varepsilon) = C\varepsilon^n$

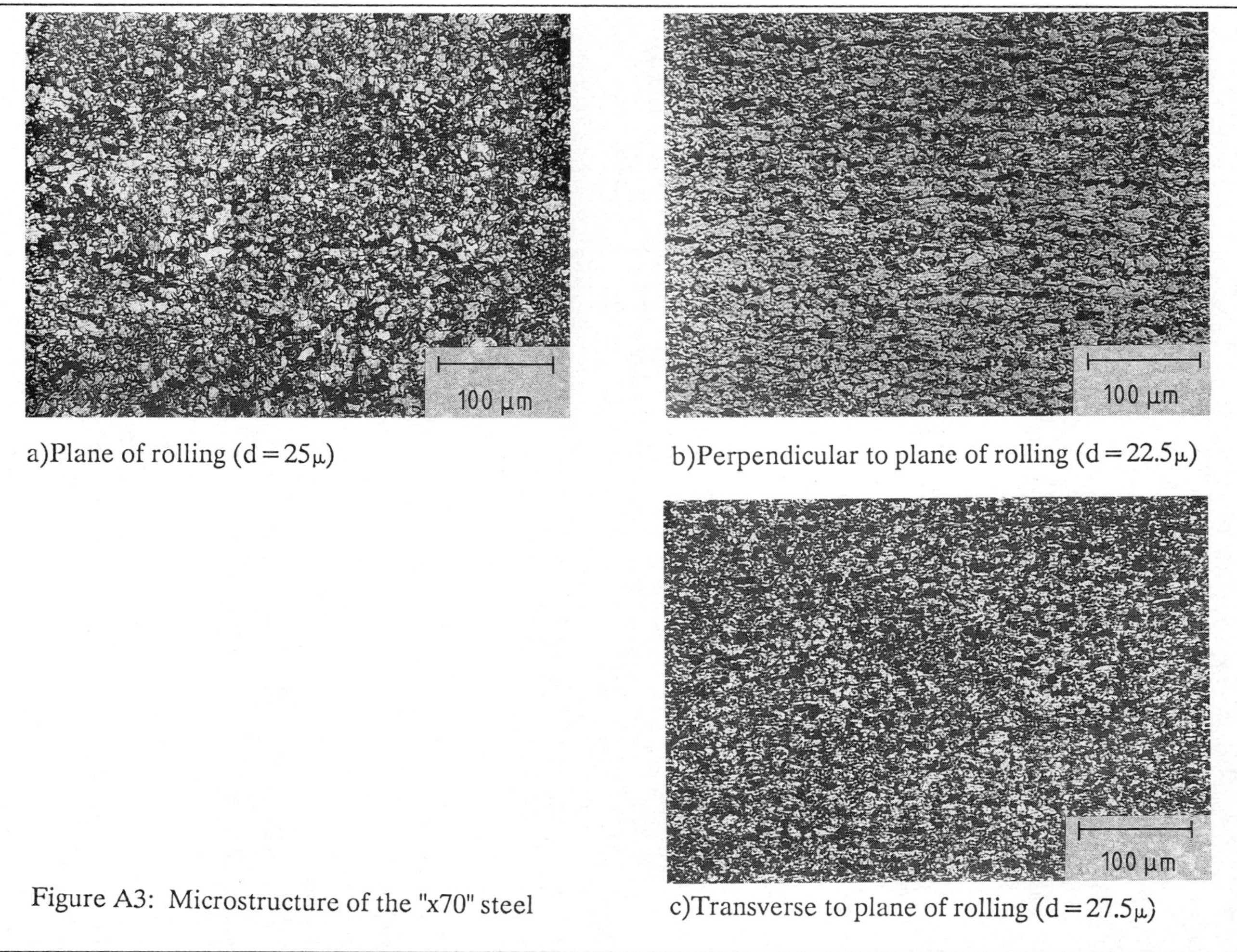

a)Plane of rolling (d=25μ)

b)Perpendicular to plane of rolling (d=22.5μ)

c)Transverse to plane of rolling (d=27.5μ)

Figure A3: Microstructure of the "x70" steel

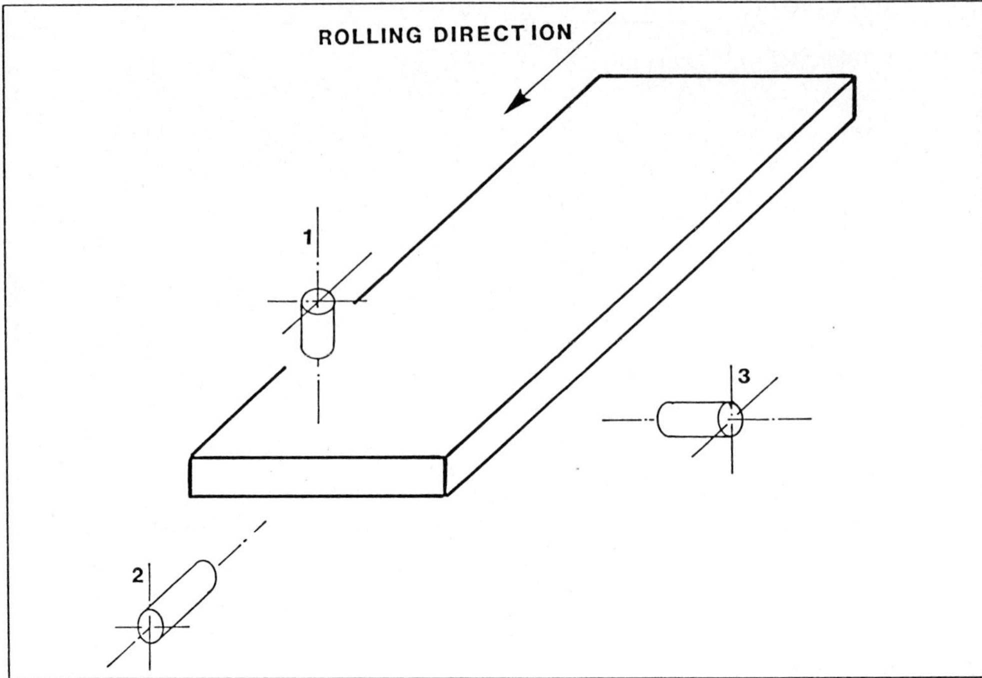

Figure A4: Specimen orientation.

when straining in excess of 90% was completed. Below that value essentially homogeneous compression took place.

Prior heat treatment

1. For samples of the first batch, cut in direction 1:
austenitizing at 1150°C for 2 minutes, air cooling to test temperature, testing and quenching in water.

2. For samples of the second batch, cut in direction 3:
austenitizing at 1150°C for 20 minutes, furnace cooling to test temperature, holding there for 5 minutes, testing and quenching in ice-brine.

Single stage flow curves

Figures A5, A6, A7 and A8 show the true stress - true strain plots obtained in uniaxial compression of samples cut in direction #1 for various constant true strain rates and temperatures. Corrections for adiabatic conditions have been included for all results obtained at $\dot{\varepsilon} = 50\ s^{-1}$.

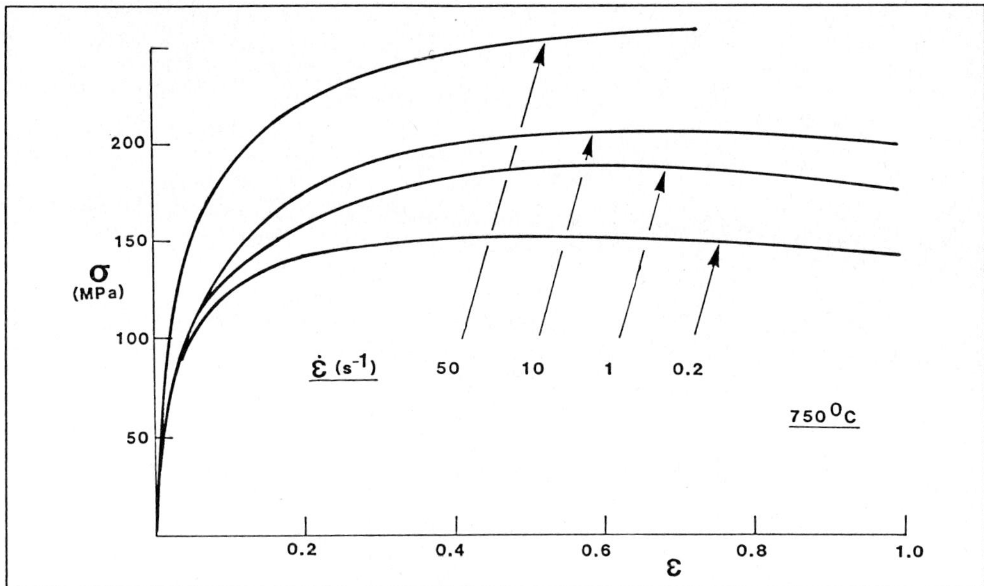

Figure A5: Flow curves as 750°C (direction #1) [13]

The formula used was

$$\Delta T = \frac{\sigma \varepsilon}{J c \rho}$$

where c is the specific heat of steel, J is the mechanical equivalent of heat and ρ is the density. Their values were taken as

$$c = 500\ J/kg°C \qquad \rho = 7850\ kg/m^3$$

Since the samples were held inside the furnace during the tests, it was assumed that all of the work was transformed into heat. It was further assumed that at $\dot{\varepsilon} = 10\ s^{-1}$ or lower, adiabatic conditions did not exist.

Peak strength - strain rate

Peak flow strength values at various strain rates and temperatures are given in Figure A9.

The strain rate hardening coefficient, n, in the Arrhenius equation

$$\dot{\varepsilon} = A\sigma^n \exp(-Q/RT)$$

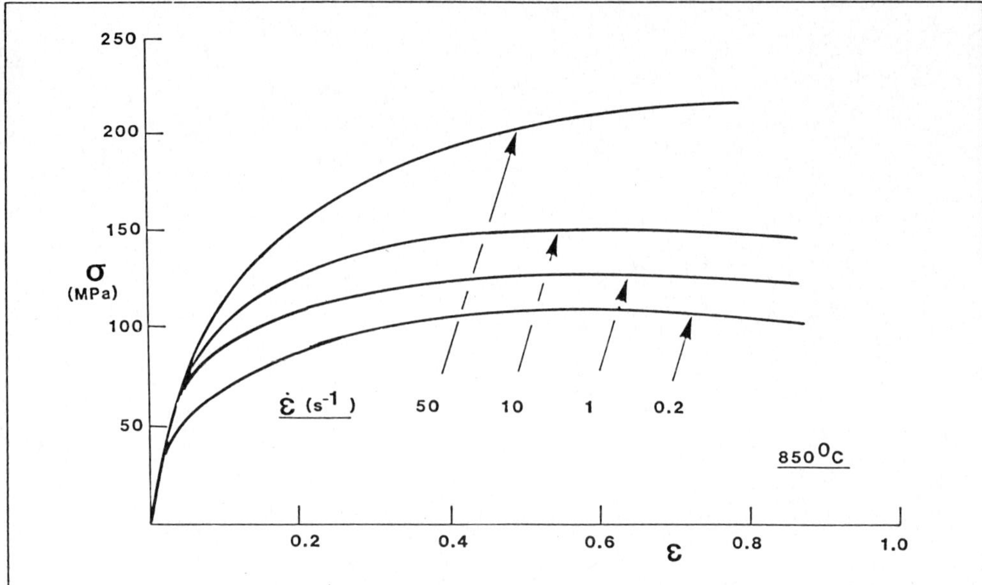

Figure A6: Flow curves at 850°C (direction #1) [13].

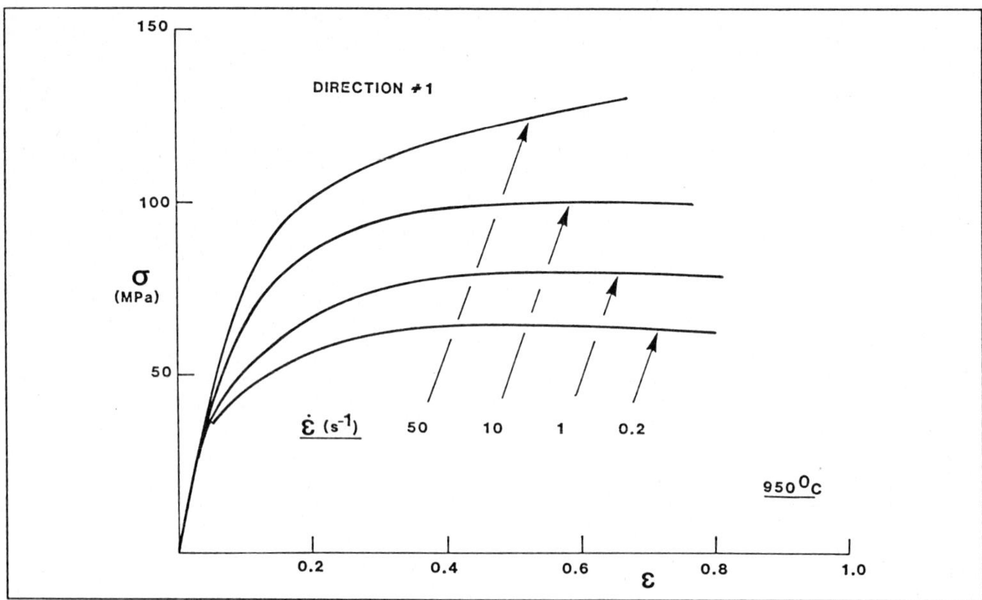

Figure A7: Flow curves at 950°C (direction #1) [13].

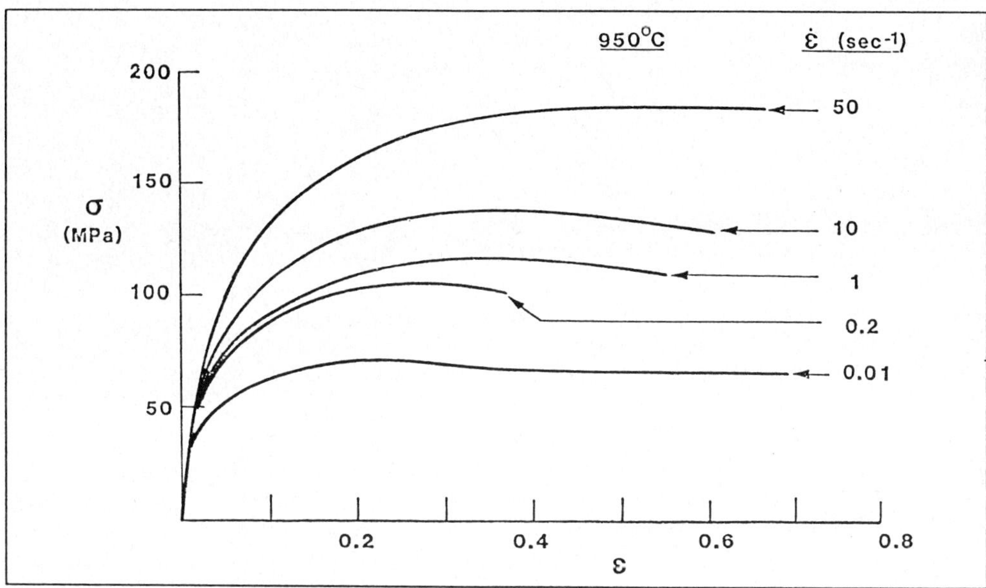

Figure A8: *Flow curves at 950°C (direction #3) [13].*

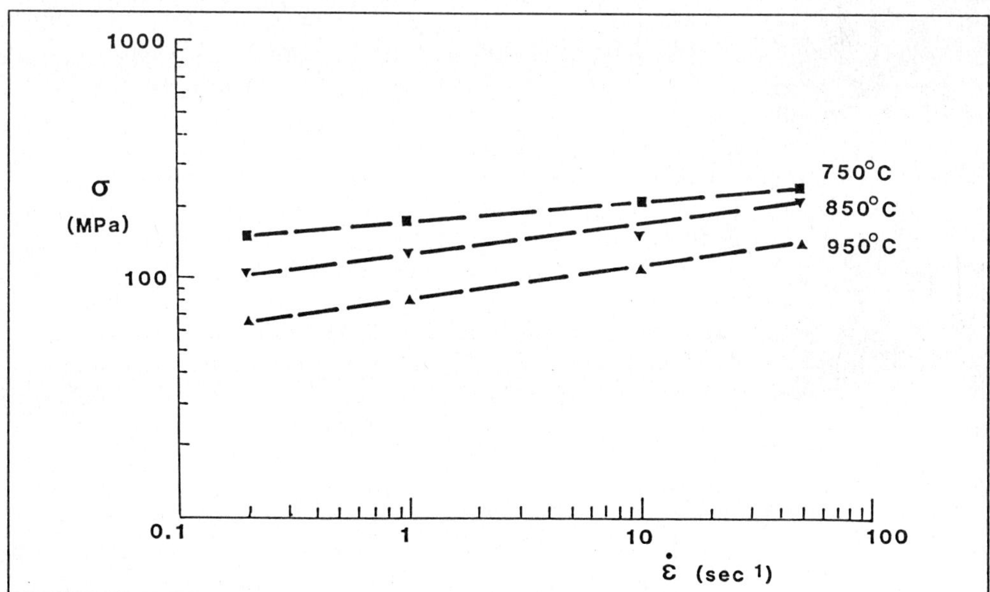

Figure A9: *Dependence of the peak strength on strain rate [13].*

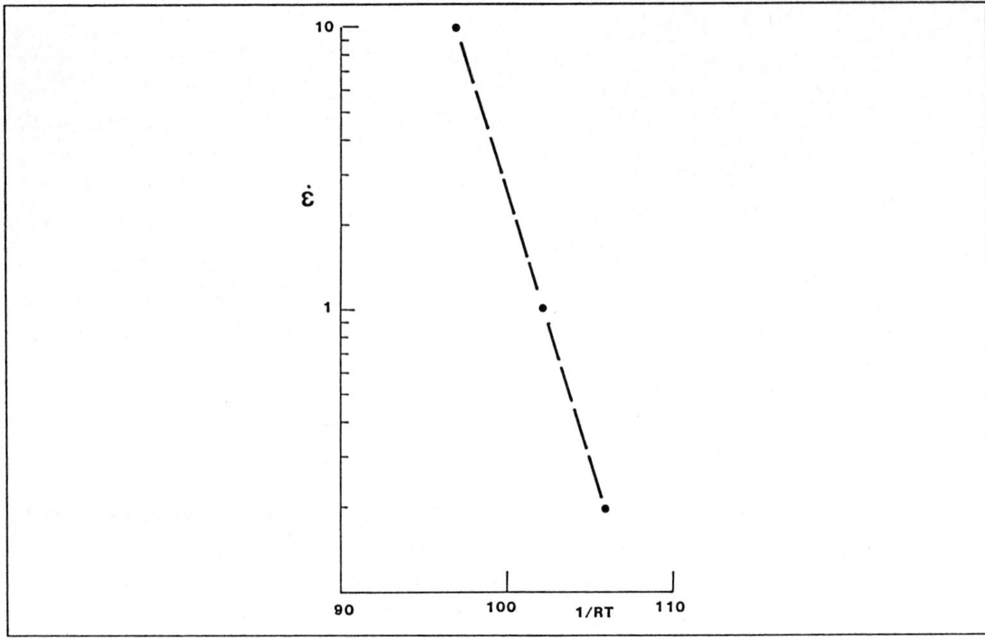

Figure A10: Dependence of the strain rate on temperature [13].

can be determined from the plot of Figure A9, using the relationship

$$n = [\ln(\dot{\varepsilon}_1/\dot{\varepsilon}_2)][\ln(\sigma_1/\sigma_2)]$$

The values thus obtained are given below:

at 950°C $n = 7.053$
at 850°C $n = 8.045$
at 750°C $n = 11.330$

SANKAR et al. [11] give $n = 8.6$; MAKI et al. [9] calculate $n = 4.61$.

Strain rate - temperature

As shown by ANAND [12], the activation energy is obtained from the slope of the $\ln\dot{\varepsilon}$ vs. 1/RT plot as

$$Q = \frac{\partial(\ln \dot{\varepsilon})}{\partial(-1/RT)}$$

From Figure A10 for a stress level of 100 MPa then

$$Q = 433.2\ kJ/mole$$

which compares well to the value given by SANKAR et al. [11] (434 kJ/mole) but not so well to the value obtained by MAKI et al. [9] (327 kJ/mole).

Microstructures

Figure A11 shows several micrographs obtained using the given process variables. Some recrystallization is evident; carbide particles are noticed. Comparing the structures obtained at 850°C, at 1 s^{-1} and at 50 s^{-1} the amount of dynamic recrystallization appears to be greater at the lower rate of strain. The microstructure of the steel, as received, is shown in Figure A12.

The effect of prior heat treatment

To test the contribution of prior heat treatment and/or anisotropy to the strength differences observed in Figures 5 to 8 three experiments were conducted - see Figure A13 for the resulting flow curves. All three tests were conducted at a strain rate of 0.01 s^{-1} and at a temperature of 950°C.

It is observed, as demonstrated by Ref. [13], that prior heat treatment has a significant effect on the uniaxial flow curves of the material. Solution treatment #1 consisted of holding the samples at 1150°C for 30 minutes and furnace cooling at a rate of 0.3°C/second to the test temperature. The sample was then held for 10 more minutes to reach equilibrium and the compression was performed. Treatment #2 included quenching after the 30 - minute hold period at 1150°C and reheating to 1050°C for 20 minutes before testing. After furnace cooling to test temperature and holding for another 10 minutes, the testing was completed. Annealing for 2 hours at 1000°C replaced the prior solution treatment for #3. The sample was then heated to 1050°C for 20 minutes, furnace cooled, held there for 10 minutes and tested.

All of the important parameters observable from the flow curves are affected by the three distinct methods of heat treatment. Peak strengths vary by as much as 10%; peak strains, indicating the beginning of dynamic recrystallization vary from 0.21 to 0.30. Evidence of cyclic recrystallization is noted to result from treatment #3 in addition to some texture induced strain hardening beyond a true strain of 0.6. Treatment #1 also shows some slow hardening while the curve for treatment #2 has reached steady state conditions at a strain level of approximately 0.75.

It may then be concluded that prior directional dependence of strength is not likely to cause major variations in behaviour. The peak strengths and strains of the annealed (treatment #3) and not annealed (treatment #2) specimens are not significantly different. As mentioned above, however, differences in prior heat treatment and in the rate of cooling to test temperature may well cause significant changes in the flow curves.

Two Stage Compression

Two experiments involving two stage compression were conducted at 750°C and the resulting flow curves are shown in Figures A14 and A15. In both cases the

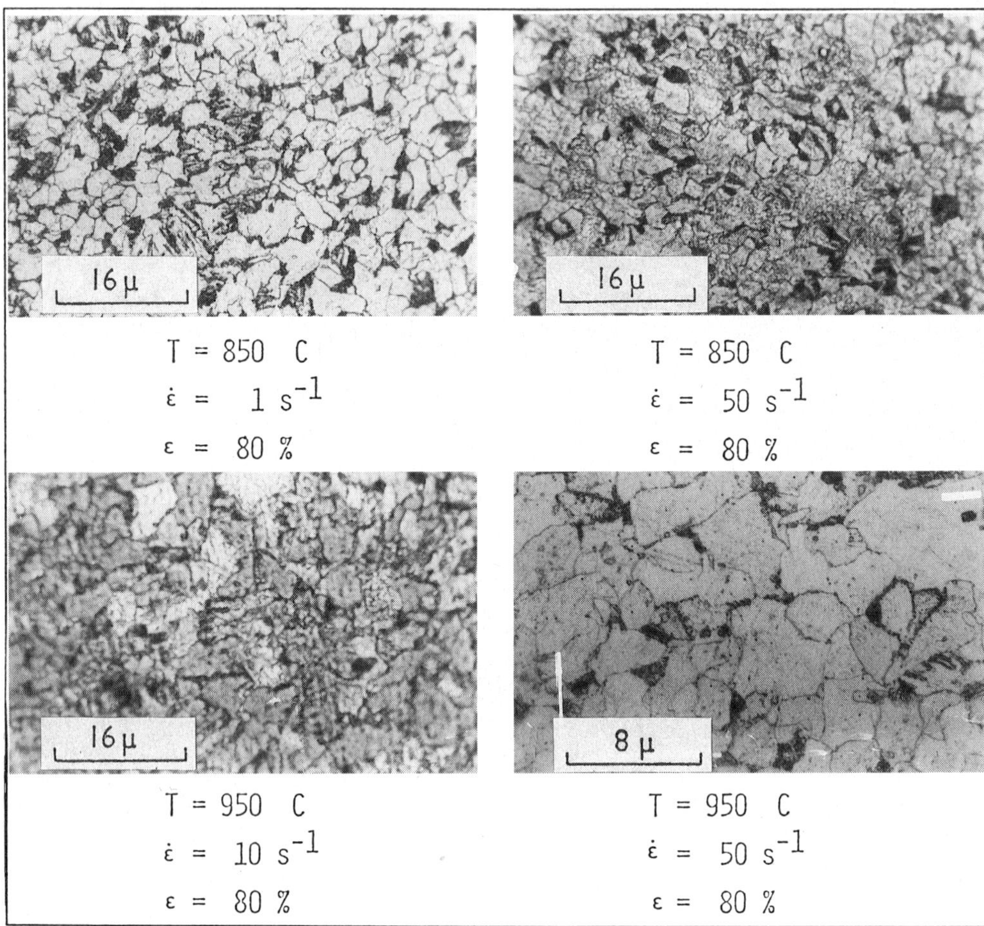

Figure A11: Micrographs of the niobium bearing steel.

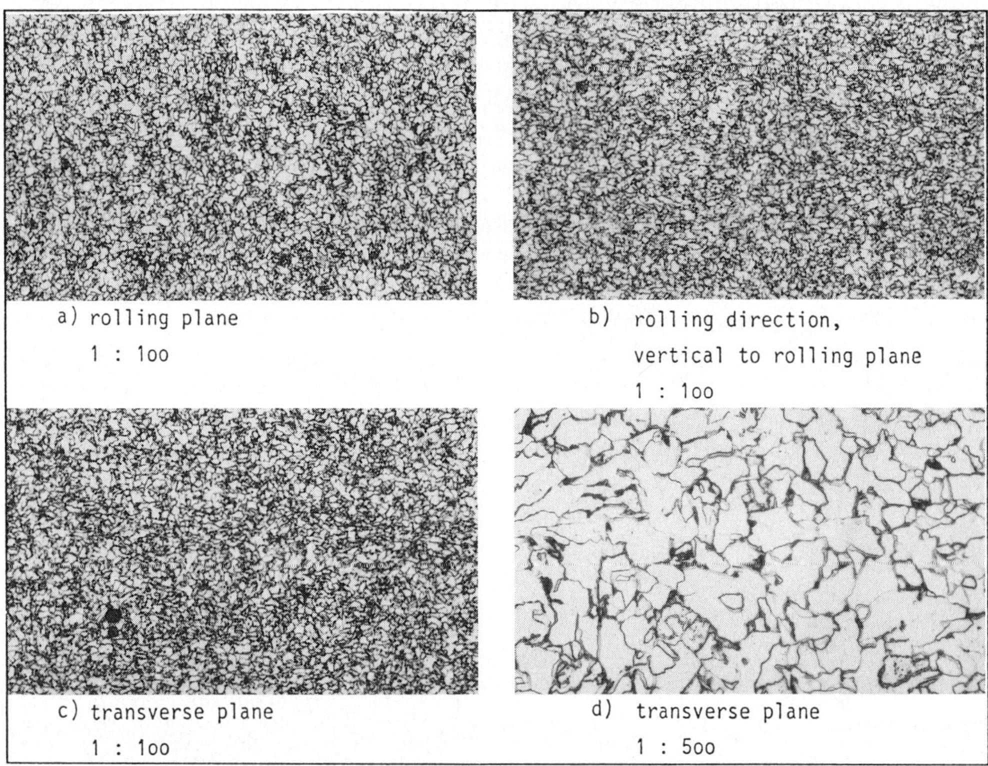

Figure A12: Micrographs of the niobium bearing steel (as received).

test was interrupted for half a second at a true strain of 0.7. In Figure A14 strain rates 5 sec^{-1} and 10 sec^{-1} were used; in Figure A15, 1 sec^{-1} and 10 sec^{-1} were employed. The flow curves appear to be practically identical; some slight difference is found in the initial portions. After interruption the material's response is unaffected by the prior mechanical treatment. The results of two stage experiments at 950°C and various rates of strain are shown in Figures A16-A19. The rates of strain are varied by a factor of one hundred and ten in Figures A16 and A17; the interruption in Figure A15 is for 1 second at a true strain of 0.45 while in Figure A16, the test is halted for 0.5 s at $\varepsilon = 0.6$. In the first cycles of loading the flow curves appear essentially identical to those obtained in single stage testing, shown in Figure A8.

During the second stages of loading, however, significant differences are noted. For a strain rate of unity, peak strain, as given in the figure is approximately 0.36. In the second stage, shown in Figure A15, $\varepsilon = 1\ s^{-1}$; no peak strain is reached and slow strain induced hardening is observed. The second stage of the flow curve given in Figure A17 is for a rate of strain of 10 s^{-1}. Again, significant hardening is observed to take place. Evidently, the first loading cycles resulted in second cycle flow curves which depended strongly on the strain rate, strain/cycle and the time of interruption and were distinctly different from the flow curves obtained in single-stage loading.

The effects of the strain at first interruption were examined next - see Figures A18 and A19. In both tests the rates of strain were taken to be 1 sec^{-1} and the test temperature was 950°C. In the first experiment the unloading occurred at a strain of 0.45 for 1 second. On reloading the material appears not to regain its strength it possessed prior to the interruption. Even on further straining to $\varepsilon = 0.6$ a drop of the flow strength of about 10 MPa is noted. As the compression proceeds further, slow rate of strain - induced hardening is displayed. When the interruption is delayed until a true strain of 0.6 - see Figure A19 - is reached, the behaviour in the second cycle changes completely. Strain hardening at a much steeper rate is found in the second cycle.

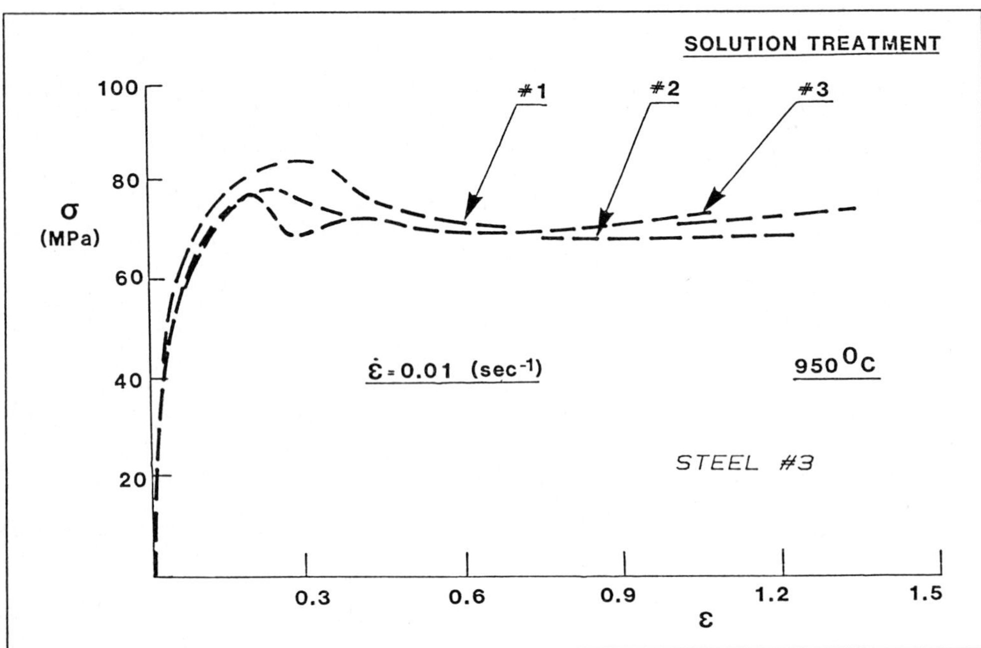

Figure A13: Effect of prior solution treatment on the flow curves (Ref. [13]).

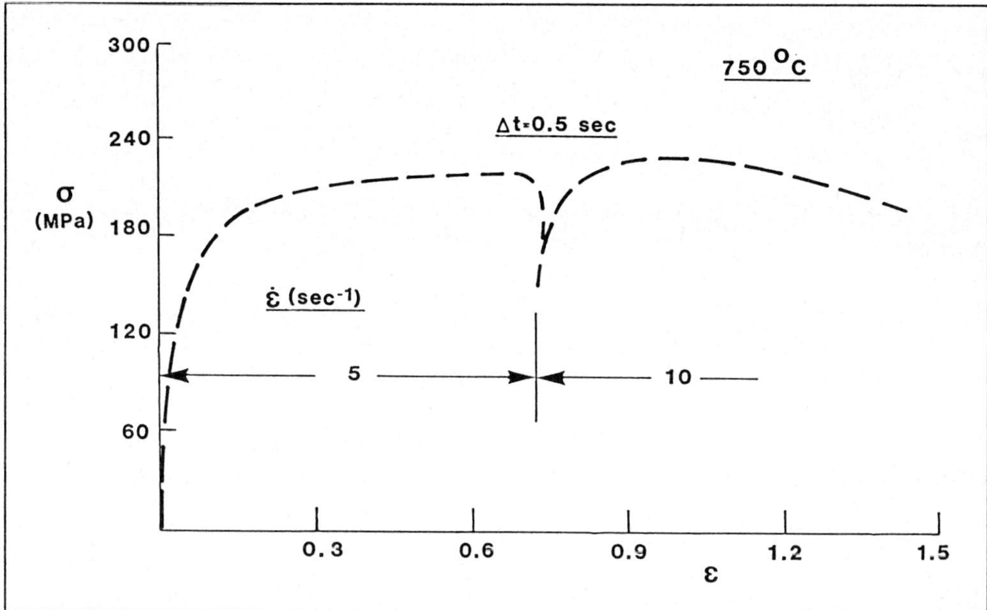

Figure A14: Two stage compression test [13].

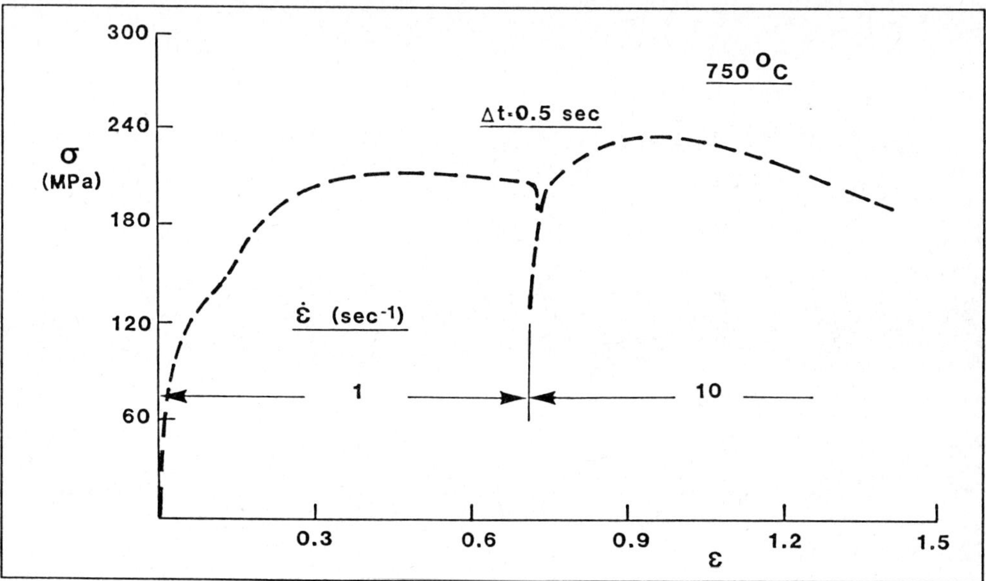

Figure A15: Two stage compression test [13].

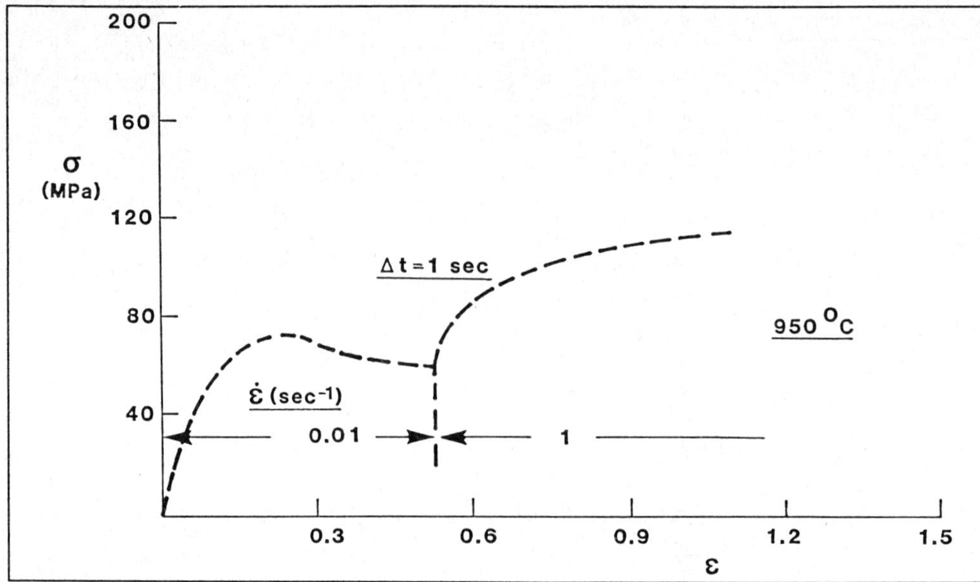

Figure A16: Two stage compression test.

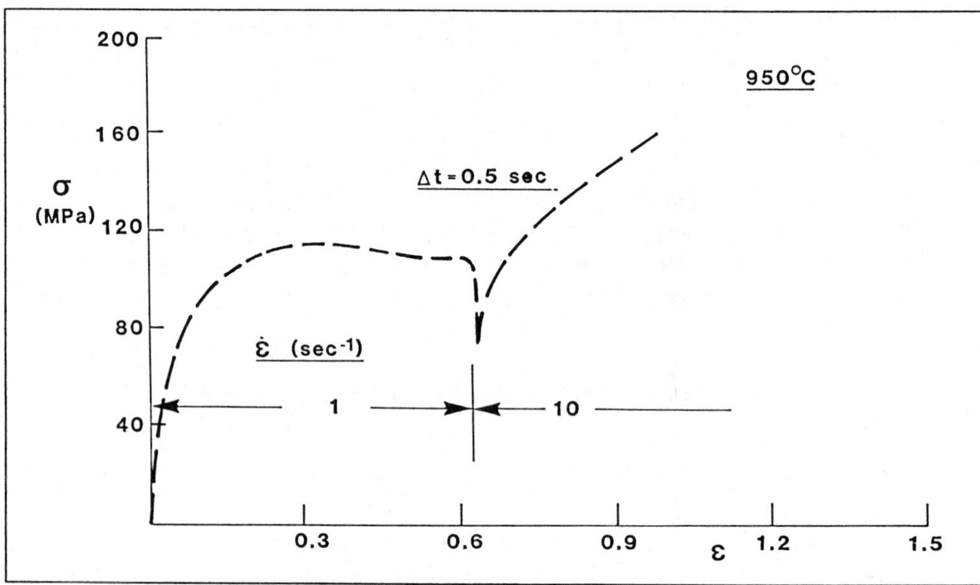

Figure A17: Two stage compression test.

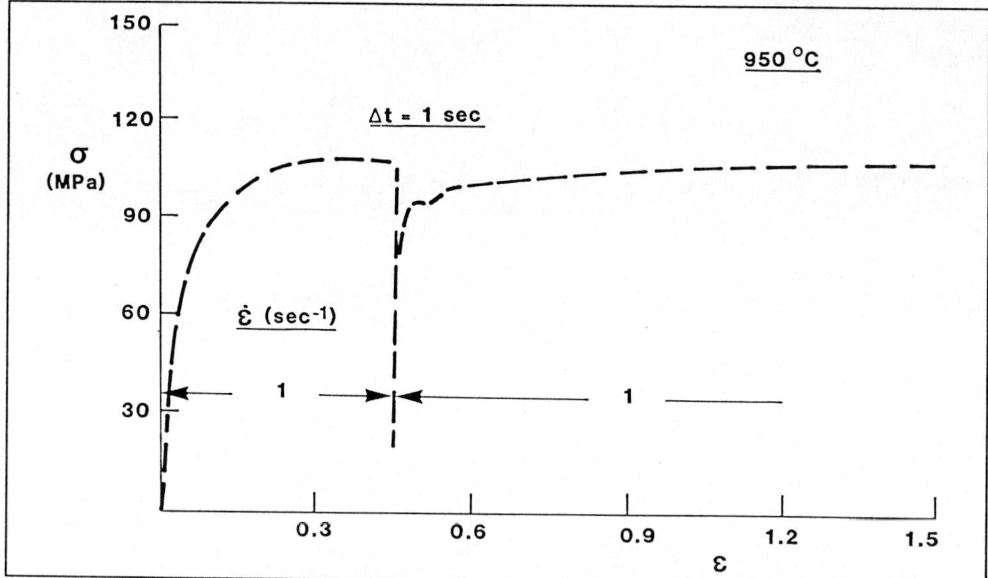

Figure A18: Two stage compression test.

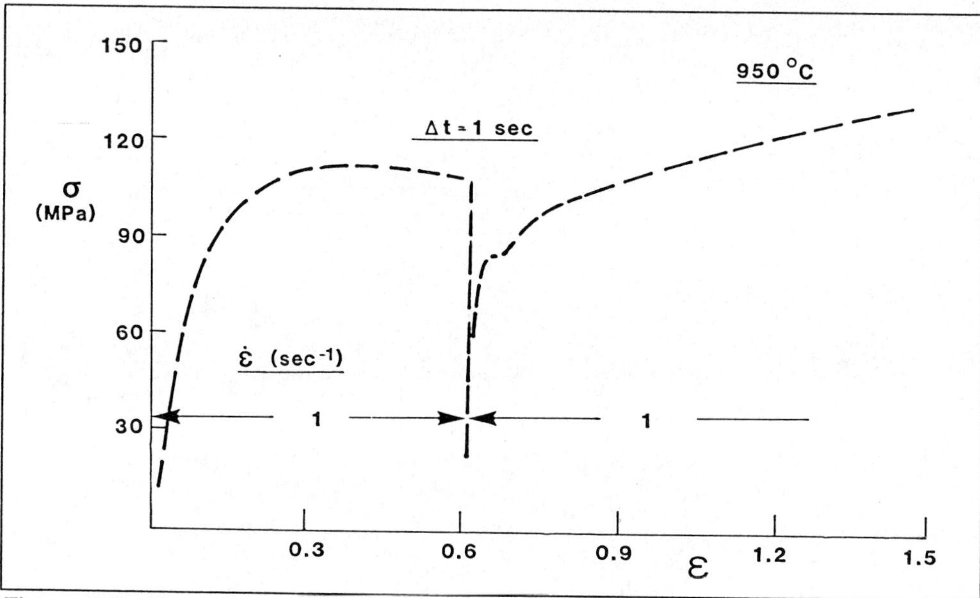

Figure A19: Two stage compression test.

4. REFERENCES

1. Shida, S., Hitachi Research Laboratory Report, 1974, 1-12

2. Altan, T., and Boulger, F.W., ASME, J. Engng. for Ind. 95, 1973, 1009-1019.

3. Gittins, A., Moller, R.H., and Everett, J.R. BHP Technical Bulletin, 18, 1974, 1-11.

4. Hajduk, M., et al., BISITS 11281, 1973, 1-10.

5. Geleji, S., "Calculation of Forces and Power Consumption During Metal Working:, II ed., Akademiai Kiado. Budapest, 1955.

6. Ekelund, S., in "Fundamentals of Rolling", Z. Wusatowski, Pergamon Press,Oxford, 1969.

7. Cornfield, G.C., and Johnson, R.J., The Journal of the Iron and Steel Institute, 211, 1973, 567-573.

8. Knudsen,W., Sankar, J.,McQueen, H.J.,Jonas ,J.J.,and Hawkins, D.U., Proc.Int. Coinf.on Hot Working and Forming Proc., Sheffield, England,1979.

9. Maki, T., Akasaka, K. and Tamura, I., Proc. Int. Conf. TPMA, Pittsburgh, 1981, 217-236.

10. Lenard, J.G., Proc. of the 3rd Seminar on Metal forming, Gyor, Hungary, 1985, 299-308.

11. Sankar, T., Hawkins, D. and McQueen, H.T., Metals Tech., 9, 1975, 325-332.

12. Anand, L., J. Engng. Mat. & Techn., 104, 1982, p. 12.

13. D'Orazio, L.R., Mitchell, A.B. and Lenard, J.G., Proc. HSLA 85, 1985, p. 281.

SUBJECT INDEX

AUTHOR INDEX

C. R. Boër,
N. M. R. S. Rebelo,
H. A. B. Rydstad,
G. Schröder

Process Modelling of Metal Forming and Thermomechanical Treatment

1986. 195 figures. XV, 410 pages.
(Materials Research and Engineering).
ISBN 3-540-16401-4

This book covers the modelling in metal forming processes, such as drawing, rolling and forging, and of thermomechanical treatment like quenching of complex parts and heat treatment of large forgings. An introduction to different modelling techniques is given with a description of the elementary analysis, upper bound analysis and finite element method applied to very large plastic deformations. Several examples from industrial practice are presented, intending to show that Process Modelling for Metal Forming have reached maturity and, in many cases, can be used to good advantage by the designer and engineers. The combination of simulation techniques, mathematical modelling and CAD/CAM technology is shown along with the advantages of their synergistic effect in CAE.

Springer-Verlag
Berlin Heidelberg New York
London Paris Tokyo Hong Kong